Tourism Crises & Destination Recovery

Sara Miller McCune founded SAGE Publishing in 1965 to support the dissemination of usable knowledge and educate a global community. SAGE publishes more than 1000 journals and over 800 new books each year, spanning a wide range of subject areas. Our growing selection of library products includes archives, data, case studies and video. SAGE remains majority owned by our founder and after her lifetime will become owned by a charitable trust that secures the company's continued independence.

Los Angeles | London | New Delhi | Singapore | Washington DC | Melbourne

SAGE

Los Angeles | London | New Delhi
Singapore | Washington DC | Melbourne

SAGE

Los Angeles | London | New Delhi
Singapore | Washington DC | Melbourne

SAGE Publications Ltd
1 Oliver's Yard
55 City Road
London EC1Y 1SP

SAGE Publications Inc.
2455 Teller Road
Thousand Oaks, California 91320

SAGE Publications India Pvt Ltd
B 1/I 1 Mohan Cooperative Industrial Area
Mathura Road
New Delhi 110 044

SAGE Publications Asia-Pacific Pte Ltd
3 Church Street
#10-04 Samsung Hub
Singapore 049483

Editor: Matthew Waters
Assistant editor: Jasleen Kaur
Production editor: Tanya Szwarnowska
Marketing manager: Lucia Sweet
Cover design: Francis Kenney
Typeset by: C&M Digitals (P) Ltd, Chennai, India
Printed in the UK

First published 2022

Library of Congress Control Number: 2021933706

British Library Cataloguing in Publication data

A catalogue record for this book is available from the British Library

ISBN 978-1-5264-0299-8
ISBN 978-1-5264-0300-1 (pbk)

This book is dedicated to the world's tourism professionals, educators and students, all of whom have been forced to confront the challenges of managing past and present crises and are hoping that this book will help them in being prepared to manage the challenges of the future.

TABLE OF CONTENTS

THE AUTHOR

Dr David Beirman, senior lecturer in tourism studies at the University of Technology Sydney, has specialised in researching destinations in crisis and disasters for over 30 years.

Dr Beirman is regularly invited to speak at major international tourism events and global tourism academic conferences. On his rare vacation time, he mows the grounds of his house in a tiny, remote Australian bush town he calls home and is currently learning the secrets of handyman jobs.

ACKNOWLEDGEMENTS

This book was inspired by the brilliant work of Joan Henderson. In 2007, she published the first thematic book on tourism crises. *Tourism Crises, Causes, Consequences and Management* set the gold standard for a thematic approach to tourism risk, crisis and recovery. Since 2007, tourism has been assailed by multiple crises, the most challenging being the onset of Covid-19. When Matthew Waters of Sage Publications approached me in late 2017 to write a book on tourism risk, crisis and recovery, I proposed an updated version of Professor Henderson's thematic approach. I sincerely thank Matthew for tolerating the many delays in writing and re-writing. Had we adhered to the original timetable and excluded Covid-19 from this book, we would have missed covering the greatest crisis affecting tourism since 1945.

I am most grateful for the 'sagacious' advice provided by Sage Editor, Jasleen Kaur, and the careful copyediting of Salina Nessa and Tanya Szwarnowska from Sage. I am also most grateful to my former student Elisa Jakymin who is now a Senior Executive with Tourism Australia; she gave me invaluable assistance with the Tourism Australia case study in Chapter 3. My wonderful researcher, Elizabeth Angus, has been a constant source of creative ideas and has an uncanny ability to locate sources I can never find.

I would like to thank my colleagues at the University of Technology Sydney (UTS). Two heads of the Management Discipline Group, Associate Professor Antoine Hermens in 2016 and Professor Carl Rhodes in 2017, supported my proposals to run two subjects on risk management at UTS. I ran my first classes on Risk and Reputation Management in Tourism in 2016 and Managing Risk and Opportunity in 2017. The latter subject has transitioned from an elective subject to a core subject from 2021 in the Bachelor of Management and the Bachelor of Business. My students in both of these subjects have been a constant source of inspiration and I am delighted that some applied the skills learned in their risk management subjects at UTS in careers following graduation. I am grateful to my Head of School, Associate Professor Bronwen Dalton, for approving my sabbatical (performance enhancement programme) during the second half of 2020. This enabled me to accelerate the completion of this book. A special thanks to Dr Steve Schweinsberg, Meg Hibbins and Barbara Almond who taught my subjects during my sabbatical period so I could complete the book free of teaching responsibilities. I have been delighted to collaborate with some of my UTS colleagues on papers and teaching relevant to this book: Professor Simon Darcy, Associate Professor Stephen Wearing, Dr Simone Faulkner, Dr Stephen Schweinsberg and Dr Najmeh Hassanli.

I'm fortunate to collaborate with some academics and tourism industry professionals in Australia, Japan, Thailand, Nepal, UK, USA, India and Israel who in their own ways have enriched the coverage of this work. I especially thank my friend and 'Master of Disaster' Dr Bert Van Walbeek with whom I worked closely on numerous tourism recovery projects and seminars in Thailand and Nepal. Bert and his wife, Gisi, are now in the UK and they have been great allies. In Nepal, I was privileged to work with Dr Pranil Upadhayaya and Mr Pankaj Pradhananga in researching and developing approaches to Nepal's tourism recovery after the 2015 earthquake. In India, Major General (retired) Ashhok Hukku has provided me with valuable insights into Covid-19 in India.

Chris Flynn (Chairman of the World Tourism Culture and Heritage Association), Anthony Cordato (Tourism Lawyer), Justin Montgomery (CEO of Amadeus Australia), Wayne Middleton (CEO Reliance Risk), Simon Elsegood (Research Director, Centre for Aviation), Rick Myatt (Chairman, Australian Travel Careers Council), and Peter Kollar (Global Training Manager, Cruise Lines International Association) are among the outstanding guest speakers who have shared their expertise in tourism risk management with my students and me over the years. I appreciate the wisdom of Dr Peter Tarlow in the US, Masato Takamatsu in Japan, Richard Gordon MBE from the UK, and Eran Ketter and Eli Avraham from Israel who appear in the pages of this book.

In Australia and New Zealand, I am grateful for the opportunity to have engaged with many fine colleagues via CAUTHE (Council of Australasian Tourism and Hospitality Educators). Since 2012, I have been the Co-Chair of CAUTHE's Tourism Risk, Crisis and Recovery special interest group with Professor Brent Ritchie of the University of Queensland and Dr Joanna Fountain of Lincoln University New Zealand. We have 40 members from ten countries all engaged in research in the field and I have been fortunate to have collaborated with many of them, including Professor Bruce Prideaux, Professor Jeff Wilks, Dr Yetta Gurtner and Dr Gabby Walters – just some of the outstanding scholars in the field.

I have always believed that tourism scholars should work closely with the travel industry. Having worked both sides of the 'town and gown divide', I've had the good fortune to work closely with the travel industry globally and in Australia over many years. I'm a founder member (since 2003) of the Australian Department of Foreign Affairs, Consular Consulting Group, a forum involving most of the Australian travel industry leadership and government working together to advance the safety of Australian travellers. My involvement has given me valuable insights into the relationship between the tourism industry and government, especially in the field of tourism risk, crisis and recovery. I am grateful for the opportunities granted me by the following international bodies to cooperate with them on a variety of projects, providing insights integral to my wider research to which I refer in this book: The Pacific Asia Travel Association,

UN World Tourism Organisation, Tourism Authority of Thailand, Nepal Tourism Board, Trinidad and Tobago Ministry of Tourism, Prefectural Government of Okinawa (Japan), Israel Ministry of Tourism, Kenyan Tourism Board, Philippine Ministry of Tourism and the Asia Pacific Economic Cooperation tourism working group and counter-terrorism working group.

David Beirman, December 2020

AUTHOR'S NOTE

Our world was never safe, yet millions of tourists over the centuries and billions since mass international travel became popular from 1970, gamely and with stout hearts, plunged into the unknown.

We're still going out there to this day, despite quake, eruption, flood, fire, epidemic, pandemic; through myriad crises and disasters, the travel industry sends them on their way and takes care of them until they return home safely. Job done.

The logistics of tourism are awesome. This multidisciplinary field is beset by seemingly insurmountable problems. It's the job of the tourism academic to lay the groundwork, a foundation from which tourism can take place, researching endlessly to provide global destinations with data and strategies to reinforce tourism operations and to keep them running smoothly, protecting the lives of tourists, the jobs of tourism staff and its phalanx – tourism training. It's an all-encompassing combined effort.

We all have to know a million things to combat the shocks of this world: whether to line up with hundreds of others to 'conquer' Mount Everest or to holiday near a potentially unstable volcano on Bali; to tour the sites of ancient civilisations amidst Middle East tensions or to brave epidemics such as cholera on the subcontinent; and to provide businesses with accurate reports on global destinations.

The information regarding decisions emanates initially from the academic sources via scouts and explorers on the ground of each destination to be scrutinised by tourism lawyers advising governments, which then release updated amendments to international tourism laws and geographic conditions. Exacting research goes into data collecting for dissemination by teams who thus instruct both the tourism industry and government advisory boards.

The world is not safe but fortunately, there are people working relentlessly to make it less threatening for billions of travellers. This book is the result of research into the Herculean efforts to sustain human interest in the impetus to explore our planet in good times and bad.

We are a living planet: stuff happens.

NB: While this book was a work in progress, the year 2020 was, for all tourism, a global crisis, a disaster that will reach into the far future as an event in which the

Earth's entire population became at risk from a virus. International travel largely ceased as travel bans restricted the movement of people.

The author's plan to spend his long-awaited sabbatical travelling the world researching was cancelled amidst the confusion, contradiction and chaos as the media sensationalised the new virus. Therefore, this book was written in situ at University of Technology Sydney (UTS) as the pandemic hysteria overtook the world. Thus, much reference to the 2020 pandemic and its effects on tourism is an eyewitness account as the book goes to press.

David Beirman, December 2020

PREFACE

INTRODUCTION

The World Travel and Tourism Council (WTTC), which closely monitors economic data for the global tourism industry, reported in 2019 that global tourism generated US$8.9 trillion in business. Tourism directly and indirectly employed 330 million people and accounted for 10.3 per cent of global gross domestic product (WTTC, 2020). The UN World Tourism Organization's (UNWTO) 2020 tourism barometer observed that in 2019, over 1.4 billion people travelled internationally (UNWTO, 2020) and many times that number domestically.

Since 1800, Europe and North America were the world's dominant tourism generating regions and principal international destinations. In 2019, France remained the world's most visited country with over 90 million international arrivals (UNWTO, 2020). However, in recent years the West's traditional domination of global tourism was rivalled by the economic rise of North-East and South-East Asia. In 2019, over 135 million Chinese citizens travelled outside China (representing almost ten per cent of global international tourism movements), positioning China as the world's leading tourism generating country. The scale and speed of tourism growth to and from the Asia Pacific is likely to result in the Asia Pacific region becoming the global epicentre of tourism activity by 2030.

By 2019, tourism became one of the world's largest industries. The growth of middle-class communities in developing countries with sufficient disposable income to travel, especially in the Asia Pacific region, coupled with the growth of low-cost travel options, appeared likely to ensure long-term future growth. From 2000 to 2019, the average year on year passenger growth of international tourism was four per cent. However, this growth rate has fluctuated due to a number of factors, including many crisis events, explored in this book.

The raw statistics reveal only a small part of the significance of tourism to the world's economy, society and environment. Tourism dominates the national economy of many countries, including developing countries. In most Caribbean countries, tourism represents more than 50 per cent of national GDP and is the main national employer. Tourism also represents close to 50 per cent of GDP in countries as varied as Fiji, the Maldives, Dubai, Oman and the Seychelles. In numerous countries, most defined as developing nations (Avraham and Ketter, 2016), tourism is widely recognised as being of national economic and strategic importance. For all countries with a viable tourism industry, tourism has represented a critical source of foreign exchange export earnings.

Tourism's importance extends well beyond the economic dimension. For many countries, tourism is integral to their national image and reputation. Consequently, real or perceived crisis events can cause significant damage to the reputation and marketability of a tourism destination or a tourism business. Crises may arise from external factors beyond destination or enterprise management control such as natural disasters, political instability, pandemics, crime, terrorism, economic slumps and widespread environmental degradation. They may also arise from issues within the control of management including negative market perceptions of unprofessional service, avaricious business practices, lack of hospitality, dishonest marketing, service or business management failures. The tourism industry is characterised by intense global competition in all its sectors. This is certainly true in the transport, accommodation, attractions, events, meetings and conventions, retail distribution and destination management sectors. One destination's crisis can easily become a neighbouring destination's opportunity.

The Covid-19 Game Changer

During 2020, a trend of long-term tourism growth since 1950 halted and dramatically reversed due to a global coronavirus outbreak, which emerged in China at the end of 2019. The Covid-19 outbreak rapidly spread around the world – in part due to the growth of international tourism. By April 2021, there were over 150 million cases registered, in 220 national jurisdictions. The virus caused just under 3.2 million deaths (WHO, 2021; Worldometer, 2020). The number affected continued to grow by over 200,000 cases daily throughout the second half of 2020 and the first third of 2021. The Covid-19 outbreak led to widespread closures of international and intranational borders and a raft of social distancing regulations designed to slow the spread of the pandemic. In addition to the significant health threat imposed by Covid-19, the pandemic's impact on global business and trade was massive. Tourism and all sectors within the global tourism industry experienced a massive decline in business and over 190 million job losses during 2020 (WTTC, 2020).

As discussed in detail in Chapter 3, the Covid-19 pandemic and the limitations to tourism imposed by a global pandemic emerged as the greatest crisis tourism has experienced since World War II. In the space of a few months, commercial airline services declined by over 90 per cent compared to their 2019 levels and international tourist visitation contracted at similar levels. During 2020, a vast number of events, cruises, flights and tours were subject to cancellation or postponement. The most notable of these was the postponement of the 2020 summer Olympics in Tokyo until 2021. The Covid-19 crisis has challenged the longstanding paradigm of tourism as an industry on an irresistible growth trajectory.

Focus on Risk, Crisis Management and Recovery

This book focuses on understanding the nature of tourism risk and crises, as they affect destinations and tourism enterprises. It analyses processes expediting the recovery of a tourism business or destination's reputation and marketability following a crisis. For destination communities and countries with a high economic dependence on tourism, the speed and quality of post-crisis recovery is often critical to the economic, political and social viability of these communities. Central to any discussion of recovery is the key concept of 'building back better' (Hollway, 2007). Effective destination and business recovery cannot be limited to merely returning to the pre-crisis status quo.

In January 2005, the UNWTO post-Indian Ocean Tsunami Emergency Recovery Conference convened in the Thai island of Phuket. At this conference, representatives from over 60 UNWTO member countries attended. Their aim was to develop a tourism recovery strategy for all destinations affected by one of the most destructive natural disasters in recent history. The tsunami killed over 275,000 people and affected Indian Ocean coastal regions in Indonesia, Thailand, Myanmar, India, Sri Lanka and even distant Kenya on the east coast of Africa (located some 8,000 km from the earthquake's epicentre off the coast of Ache Province, Indonesia). In Phuket, it was immediately evident to all delegates that the normally thriving tourism businesses in Phuket and tourism dependent businesses such as markets and restaurants had halted. At least half of the casualties from the Indian Ocean tsunami in Thailand's Andaman Sea coastal regions were international tourists. For the people of Phuket, one of Thailand's most popular tourism destinations with an economy largely dependent on tourism, the recovery of tourism was a matter of economic survival. The Tourism Authority of Thailand has responded to a variety of crisis events in recent years and quickly proved its effectiveness in restoring tourism confidence in, and demand to, Phuket (Beirman, 2018; Gurtner, 2007).

A concept closely associated with post-crisis tourism recovery is resilience. Central to resilience is the capacity, strategic approach and will of a tourism business or destination to recover rapidly from setbacks. Some crisis events, especially one-off, aberrant events such as a natural disaster, present an opportunity for rapid recovery. In such cases, volunteer or solidarity tourism can be an integral part of the recovery process. The author was involved as a member of the Pacific Asia Travel Association's Nepal

Rapid Response Taskforce, which joined forces with Nepal's tourism industry leaders to develop a strategy for Nepal's tourism recovery after the destructive April 2015 earthquake. Nepal's rapid recovery from this disaster was facilitated by a significant growth in volunteer tourism and an emphasis on well-targeted, low-cost e-marketing (Wearing et al., 2020.

Conversely, long running crises, which undermine a destination's reputation, may involve a prolonged recovery process. Israel's tourism recovery after the Al Aqsa Intifada 2000–2004 took several years (Beirman, 2009). In fact, many Arab countries, including Egypt, Tunisia, Libya and Yemen, which experienced political uprisings and ongoing political instability since the beginning of the 'Arab Spring' political uprisings in 2011, have shared the realisation that restoring the confidence of tourists to return to destinations which have experienced long-term social and political unrest is a slow and arduous process. In April 2019, a civil protest movement that opposed the local government in Hong Kong had escalated by November 2019 into wide-scale violence. The ongoing political unrest has transformed the tourism and business image of Hong Kong from one of the world's most desirable destinations to a danger zone for tourists and traders alike.

Tourism Risk, Crisis and Recovery Study and Research

In 2016, the author commenced teaching the subject Reputation and Risk Management in Tourism at the University of Technology Sydney (UTS). In 2017, he developed, introduced and taught the subject Managing Risk and Opportunity, a generic risk management subject for business students focusing on enterprise recovery. UTS is not unique in offering subjects or courses that deal with tourism risk, crisis and recovery issues. A growing number of universities, research centres and institutes of higher learning around the world are devoting increasing research and teaching resources to the critical issues of risk, crisis and recovery management in tourism. The teaching and research of tourism risk and crisis management occurs at a number of universities and colleges around the world. These include: University of Bournemouth, University of West London, Oxford Brookes University, Leeds University, Strathclyde University (UK), University of Haifa (Israel), Central Florida University, University of Southern Florida, University of Hawaii (USA), Hong Kong Polytechnic (Hong Kong), James Cook University (Queensland, Australia), Taylor's University (Malaysia), Dhurakij Pundit University (Bangkok, Thailand), Mae Fah Luang University (Chiang Rai, Thailand), and Stenden University (South Africa). Many institutions teach subjects and host research centres in the field of tourism risk, crisis and recovery management for their academics, undergraduate and postgraduate students, and for travel industry leaders and senior professionals.

The Council of Australasian Tourism and Hospitality Educators (CAUTHE), which represents tourism and hospitality scholars and researchers in Australia and New Zealand, has several special interest groups focusing various fields of tourism research, including sustainability, volunteer tourism and other significant areas of research. By February 2020, one of the largest special interest groups in CAUTHE was the Tourism Risk, Crisis and Recovery special interest group. Other international groups of tourism academics, including the BEST Education Network, have featured tourism risk, crisis and recovery management as a primary theme at their annual conferences.

The output of academic and professional literature on tourism crisis and destination recovery has undergone exponential growth since 2001. The terrorist attacks that destroyed New York's World Trade Center and severely damaged the Pentagon in Washington with a loss of almost 3,000 lives on 11 September 2001 was the traumatic catalyst for tourism crisis management to undergo a rapid transformation, from a peripheral and niche area of study and practice in tourism management, to a core issue.

Tourism Industry Recognition of Tourism Risk, Crisis and Recovery

The 9/11 terrorist attack and the political and US-led military response to it continue to have widespread global geopolitical and economic impacts, especially in Western countries and the Islamic world. The global tourism industry was one of the first business sectors to feel the economic impact of 9/11. In the six months after 9/11, there was a significant global decline in commercial airline patronage. New York City and the remainder of the USA experienced a serious decline in international tourism arrivals, which took several years to recover. However, the US was one of many destinations in the world in which tourism flows declined following the attack that primarily targeted New York City and Washington, DC.

In the months following 9/11, the global airline sector, led by the International Air Transport Association, Airports Council International and the International Civil Aviation Organisation, undertook the task to review global airline and airport security measures. The 9/11 attack had magnified the longstanding threat of airline hijacking into a far more dangerous threat of commercial airlines being employed as weapons of mass destruction. This security review took place during late 2001 and 2002. After the 9/11 attack, the world's airlines and airports undertook the process of reassuring all their business stakeholders and the global travelling public that the aviation sector had effectively addressed the security failures that enabled the 9/11 hijackings to occur. The airline and airport sectors have traditionally been the most proactive sectors in addressing risk in global tourism. They are also subject to the strictest sectoral global governance in global tourism.

The International Civil Aviation Organization (ICAO) as a United Nations (UN) body has the power to impose safety and maintenance standards on every airline in the world. Consequently, the airline sector was in a very strong position to respond in a unified global manner to a sectoral security crisis. The International Maritime Organisation (IMO), the UN's governance body for global shipping, has established strict standards for risk management, embraced by the global cruising sector.

Other sectors of the global tourism industry lack the degree of global governance of the airline, airport and cruising sectors. Consequently, responses by other sectors such as the destination management, accommodation, events and attractions sectors to risk and security threats tend to be far less globally consistent than airlines, airports and the cruising sector.

The highly diverse accommodation sector has demonstrated an ongoing problem in responding to security threats. This is largely due to its seemingly contradictory role as a place of 'hospitality' and the open access this implies, countered by its duty of care for the safety and security of its guests. The Las Vegas massacre of 1 October 2017 in which a gunman used his hotel room as a firing nest to kill 58 and wound over 500 people attending a concert next to the hotel raised many questions about luggage screening, which many hoteliers have been reluctant to address. The Islamist attack on three luxury hotels in Sri Lanka on 21 April 2019 involved three suicide bombers (registered hotels guests) who came to breakfast and detonated their body bomb devices, killing 39 hotel guests and wounding many more.

The 9/11 attack forced the tourism industry beyond the airline sector to take risk and crisis management seriously. Since September 2001, multiple events which negatively impacted on the desirability of tourism destinations forced the travel industry to understand that risk and crisis management was not the exclusive domain of airlines. The October 2002 Bali bomb attack, targeting two nightclubs in Denpasar, killed over 200 people (mainly tourists). The SARS pandemic scare of 2003 severely reduced tourism in South-East Asia and Hong Kong during the first half of 2003. The Madrid Railway Station attack and the Indian Ocean tsunami in 2004 were among a succession of events negatively affecting the perception of tourism destinations. Consequently, all sectors of the travel industry were obliged to realise that risk and recovery management applies to all sectors of the tourism industry, and could occur in any destination region in the world.

In recent years (2015–2019), terrorism targeting tourists in the UK, France, Belgium, Turkey, Russia, Tunisia, Spain, Philippines, US, Egypt, Israel, Thailand, Sri Lanka and Kenya (primarily acts of terrorism undertaken by jihadist groups) has heightened global concern about the impact of terrorism and the cycle of violence on tourism. A detailed discussion of terrorism is in Chapter 5. However, it is important to note that the addiction to global publicity on the part of most terrorist groups makes international tourism and tourists especially compelling targets (Tarlow 2014a).

Crime, pandemics such as the Zika virus (which now exists in over 100 countries), Ebola (largely confined to West Africa), and the 2020 Covid-19 outbreak and political instability have been extensively reported by the international media, on a growing number of instantaneous globally-distributed platforms. The role of globalised media and the expansion of mobile technology and social media platforms are increasingly important today, as the dissemination of negative news is instantaneous and global.

The challenge of recovering visitation to tourism destinations and businesses after a wave of negative publicity requires tourism managers and marketing professionals to undertake training, enabling them to acquire a mastery of communicating recovery messages, utilising those media and IT platforms.

The vulnerability of tourism to crisis events is both actual and perceptual. The Syrian civil war since 2011 largely destroyed Syria's formerly vibrant tourism industry. In 2010, Syria attracted almost ten million international visitors. By the middle of 2020, the Syrian civil war caused the death of at least 500,000 Syrians, rendered millions more homeless and inflicted incalculable property and infrastructure damage. Syria represented a clear case in which tourism had become all but unviable. However, this did not deter Syria's Assad regime in March 2016 from attempting a surreal YouTube advertising campaign, 'Syria Always Beautiful' to convince the world that Syria is a desirable destination for tourists (Becken and Carmigani, 2016). The Syrian civil war was a compelling and finite cause for a major tourism downturn.

By contrast, the downturn of tourism in neighbouring Jordan between 2011 and 2017 was largely perceptual. Despite the fact that between 2011 and 2017 Jordan's neighbouring countries, Iraq, Syria, Egypt and Israel experienced events that compromised their reputation as safe destinations, Jordan was relatively free of internal political violence. However, the widespread perception of Jordan as an integral part of a politically unstable region has led to a decline in international tourism to Jordan during this period (Beirman and Walters, 2017).

In November 2017, the Bali Tourism Board's delayed response to the eruption of the Mount Agung volcano severely impacted on Bali's reputation as a tourism destination. The volcano's eruption and ash cloud caused no tourism casualties. However, the mixed messages tourists received from airlines, insurance companies and Balinese tourism authorities resulted in a perceptual crisis. The perception of crisis was dominated by extensive media coverage of 'distressed, stranded and ripped off' tourists whose Balinese holiday was disrupted by flight cancellations due to airlines' legitimate concerns about avoiding volcanic ash clouds, emanating from Mount Agung.

Students, researchers and practitioners of tourism crisis and recovery management should be aware that crises arise from both actual and perceptual stimuli. Consequently, the recovery of tourism requires addressing both the finite and perceptual dimensions of a crisis. The challenge for destinations and tourism businesses involves the effective and timely communication of transparent and verifiable facts to overcome perception and speculation. This skill applies to both traditional media platforms (radio, TV and print) and the wide range of increasingly influential social media platforms.

The global tourism industry is gradually increasing its attention to crisis and recovery management. The UNWTO has a well-established risk and crisis management unit since 2004. At the WTTC Global Tourism summit, held in Bangkok in April 2017, the WTTC declared that tourism crisis management along with sustainability and technological development were its three core challenges for global tourism. In July 2020, in response to Covid-19, the WTTC released its safe tourism protocols (WTTC, 2020), which applied to all sectors of the tourism industry. The UNWTO, Pacific-Asia Travel Association (PATA), ASEAN Tourism, the European Tourism Association, the Caribbean Tourism Association and Asia Pacific Economic Cooperation Tourism and Counter terrorism working groups are just a few of the transnational tourism and government associations that have developed strategies to address crisis management, security, destination and business recovery. Increasingly, national, state and provincial tourism associations have developed crisis management strategies, guides and training modules for their tourism industry stakeholders.

The output of books, journal articles, doctoral and masters by research dissertations and research briefs by tourism academics and higher degree research students has grown significantly since 2001. The author's first book, *Restoring Tourism Destination in Crisis – A Strategic Marketing Approach* (Beirman, 2003) was one of a handful of books addressing tourism crisis and recovery. Since then, global academic output in this field has resulted in the publication of over 100 books, thousands of journal articles, conference papers and research briefs. Tourism risk and crisis management is frequently addressed at both academic and tourism industry conferences all over the world.

This book refers to the growing body of literature involving many distinguished scholars who have contributed to the growing body of knowledge in this field. However, as most tourism scholars and practitioners would readily agree, there remain many significant research gaps and opportunities for advancing research in the field.

The approach of this book involves a thematic focus on the recovery of tourism businesses and destinations. Very few books in the field of tourism crisis and recovery management (especially those written by a single author) have attempted to approach tourism crisis management thematically. The most significant exception was *Tourism Crises: Causes Consequences and Management* written by Joan C. Henderson (2007). Henderson's book has served as an excellent introductory textbook for tourism students and professionals seeking an introduction to tourism crisis typologies and management approaches to them. Since 2007, many subsequent crises have profoundly affected tourism. Recent advances in IT technology, 24/7 news media coverage and online platforms require a fresh examination of the themes covered so well by Joan Henderson in her book.

Additionally, since 2007 the global tourism industry has played an increasingly active role in the fields of proactive and reactive tourism risk, crisis and recovery management. Although the primary focus of this book is destination recovery, it's appropriate that tourism enterprises and key sectors of the tourism industry are also examined to provide a broader context.

Many cases referred to in this book to illustrate the themes arise from the author's first hand research and direct involvement.

THE MAIN THEMES COVERED IN THIS BOOK

Part 1: Establishing the Theoretical and Sectoral Context for Tourism Risk, Crisis and Recovery with focus on Covid-19 Tourism's most Challenging Crisis

Chapter 1 – A Contextual Approach to Tourism Risk, Crisis, Recovery and Resilience

This chapter covers key definitions, examines the academic literature on tourism crisis and recovery, and discusses the application of collaborative theory that's central to effective tourism recovery. It also discusses the issue of tourism resilience, which has emerged as a major area of research and practice in global tourism. Central to the chapter is a discussion of the core elements of risk management with a focus on identifying likelihood and consequence in the strategic management of risk and crisis management. This chapter will also address the issue that many destinations and tourism businesses can confront multiple risks.

Chapter 2 – Government and Travel Industry Approaches to Tourism Recovery

This chapter focuses on tourism industry and governmental approaches to destination and tourism sectoral recovery. In recent years, until the onset of Covid-19, the global tourism industry has made significant advances in addressing tourism risk crisis and recovery. The chapter discusses tourism destination recovery programmes developed by the UNWTO, WTTC, APEC, PATA and some national and regional tourism organisations. It also discusses sectoral approaches undertaken by the airline, cruise, accommodation and wholesale tour operator sectors. Government has played a crucial role in financially assisting tourism enterprises and destination marketing organisations in riding out the massive financial impact of Covid-19 during 2020.

Chapter 3 – Unprepared: The Global Tyranny of Covid-19

This chapter examines the massive impact the 2020 Covid-19 pandemic has had on global tourism, the lack of preparedness for a global pandemic on this scale, and the response of the tourism industry and governments. It is clear that Covid-19 has emerged as a watershed event for global tourism, and the chapter focuses on its impact on established paradigms of global tourism and, specifically, approaches to risk, crisis and recovery management.

Covid-19 has proven to be far more than a flu-like pandemic, infecting millions and killing well over a million people. In the space of a few months from January 2020, the novel coronavirus (Covid-19), which apparently originated in China, spread globally and instigated the most far-reaching global changes to travel and tourism governance of any event since the invention of powered aircraft.

The common 20th and 21st century paradigms of minimally restricted global travel, the social democratisation of tourism and continuous tourism growth have been debunked. During 2020, most countries and many government jurisdictions within countries suddenly closed their borders or restricted access to outbound and inbound travellers. Measures to limit the spread of the virus involved social distancing regulations in virtually every country in the world. In turn, the pandemic altered the way tourism business operates in every sector of the tourism and hospitality industry.

One of the distinguishing factors of the Covid-19 pandemic has been its destructive and disruptive impact on every sector of the tourism industry. The pandemic has enforced heavy reliance of many tourism enterprises on government support, for businesses and their employees to survive a crisis, which denied large sections of the tourism industry of their core business.

Part 2: Key Themes in Tourism Risk, Crisis and Recovery

Chapter 4 – Theme 1: Political Instability and Tourism

There is a close relationship between political instability and terrorism. However, not all episodes of political instability necessarily generate terrorism and especially terrorism targeted at tourists. The bout of civil unrest in Thailand between 2009 and 2011 and the Turkish government's response to the failed coup attempt of July 2016 are two examples in which political uncertainty has damaged a destination's image and reputation. However, the extent and severity of political instability will determine the nature of tourism's response to it. There can be significant differences. There are uncertain tourism implications of Britain's commitment to Brexit and the economic problems of Greece. But in both cases, tourism effectively operates. By contrast, the reduction of countries, including Syria, Somalia, Yemen, Libya, Venezuela and Iraq, to the status of 'failed states' led tourism to become unviable.

Chapter 5 – Theme 2: Tourism Recovery from Terrorism

This chapter examines recent outbreaks of terrorism, focusing on terrorist attacks targeting tourists. It discusses counter-terrorism and tourism, and post-incident recovery approaches. Some of these strategies were formulated in 2017 during the APEC Counter-Terrorism Working Group, which held a conference on terrorism and its impact on tourism in Bali in May 2017. It will also examine the approaches taken by the airline and airport sectors in response to the 9/11 attack and discuss the challenges to more vulnerable sectors of the tourism industry, including the global accommodation, attractions and events sectors. The discussion on terrorism includes reference to the 'isolation' strategy employed by many destinations, recovering confidence in source markets to distinguish between danger zones and the broader context of safety.

Chapter 6 – Theme 3: Destination Recovery from Natural Disaster

This chapter examines tourism crisis management and destination recovery approaches at the transnational, national and local levels. Among cases examined are the recovery from the 2011 Japan earthquake/tsunami, the 2004 Indian Ocean tsunami, Samoa's 2009 tsunami, the Nepal earthquake of 2015, Cyclone Pam, Vanuatu, 2015, Victorian bushfires of 2009 and the Australian bushfires of 2019–20. The chapter refers to the UNWTO's collaborative crisis management approach, developed in successive conferences held in Australia in 2011 and 2012. These conferences focused on the integration of tourism enterprises and associations into the broader strategic dimension of emergency management.

Chapter 7 – Theme 4: Health-Related Tourism Crises with an Emphasis on Pandemics

This chapter focuses on both real and perceptual medical crises. Media coverage of the 2003 outbreak of SARS in North-East and South-East Asia instilled a level of fear among travellers, far greater than the actual extent and impact of the disease. This chapter discusses the contextual background of health-related issues and assesses strategic and operational approaches to destination recovery from the impacts of a range of recent medical concerns, including H1NI (swine flu) in 2009–2010, MERS in 2015, Ebola in 2015 and the Zika virus, which gained attention in 2016 in advance of the Rio Olympics. Although a separate chapter is devoted to the 2020 Covid-19 outbreak, reference is made here to Covid-19 within the broader context of pandemics, with a major impact on tourism.

Chapter 8 – Theme 5: Crime and Tourism

The prevalence of crime proved to be a deterrent against visitation to many tourism destinations in the world. During and prior to the 2016 Rio Olympics, extensive media focus on crime in Brazil ranging from assault to opportunistic theft and scams deterred many international spectators and some athletes from attending. This chapter examines both proactive preventative measures, including Tarlow's (2014b) Tourism Orientated Police and Protection Services (TOPPS) programme, Bali's community policing and tourism recovery strategies successfully practised by tourism destination authorities in South Africa and the Caribbean. While government travel advisories from tourism generating countries warn their citizens about crime risks in destination countries, relatively few countries provide pre-arrival warnings of common crime threats to incoming tourists.

Chapter 9 – Theme 6: Economic Crises

The 2008–2009 global financial crisis (GFC) led to a significant downturn in global tourism demand. One of the lessons many destinations learned from the GFC was the importance of having a diverse tourism infrastructure and product offering, which appeals to both large and small travel budgets. Destinations, which limited their tourism offering to the premium sector of the market, suffered the greatest loss of tourism during the GFC. Marketing and product flexibility is one of the key recovery approaches to managing economic downturns.

More recently, the Covid-19 pandemic of 2020 has created a global economic crisis, which has impacted on all sectors and levels of business, with tourism one of the most heavily affected sectors. Business travel was rapidly replaced by online business meetings and transactions, which now present a significant challenge to the future recovery of business travel.

Chapter 10 – Theme 7: Technological Risk, Crises and Recovery

Airlines, cruise ships and high-speed rail are generally considered as safe modes of transport, but when disasters occur (air crashes, ships sinking or train derailments) they tend to cause high casualties and can severely damage the reputation of the operators or the brand.

Loss of power or power surges have the potential to endanger operations at major events, attractions and guests in hotels.

Today's high level of reliance on technology for tourism booking and reservation transactions, financial transactions and accommodation, attraction and transport facilities opens the question of how destinations and businesses can respond to cybercrime and data loss. What risk management measures are employed if and when computers are hacked, the power is switched off, or the technology that travellers and tourism professionals rely on in all sectors of the travel industry ceases to work?

Chapter 11 – Theme 8: Recovery from Internal Management, Product and Service Failures

While many of the crises discussed in this book result from external causes, many destinations and tourism enterprises are the makers of their own misfortune.

Costa Cruises failed to respond appropriately to the 2012 Costa Concordia disaster, in which the ship sank just metres from an Italian island (killing 35 passengers). The same company responded poorly to Costa Allegra's engine malfunction and power loss in the Indian Ocean weeks later, which transformed a deluxe air-conditioned voyage to the Seychelles into a 38°C voyage from hell for several hundred wealthy passengers, leading to significant reputational and brand damage and massive legal bills for Costa Cruising.

In 2017, United Airlines forcibly removed a passenger (thought to be of Chinese extraction but actually of Vietnamese background) from one of its domestic aircraft in the US, due to overbooking. This action led to well over 100 million people viewing the incident on a variety of social media platforms in China alone within 24 hours of the incident. United's poorly worded response to this incident, which they dismissed as 're-accommodation', led to a significant drop in United's share price and widespread calls in the US and worldwide for boycotts of the carrier.

In September 2019, the financial collapse of one of the world's most iconic tourism companies, Thomas Cook, had a major impact on British and European tourism. On a single day (September 23, 2019) 600,000 predominantly British and European passengers faced massive disruption to their holiday arrangements; 22,000 Thomas Cook employees were jobless and thousands of tourism businesses that had accepted Thomas Cook bookings for tour and accommodation services found these bookings were unpaid. As a vertically integrated travel company, Thomas Cook operated an airline, owned hotels, resorts and coaches, and ran travel agencies and tour operator offices. All these enterprises ceased operation simultaneously. This was a consequence of failing to manage a debt, growing since 2007, and which by September 2019 had reached over US$3 billion. Thomas Cook's senior management had long been aware of the company's deteriorating financial position. However, until the collapse of the company, the extent of the problem was unknown to its stakeholders. Its impact on employees, customers, business stakeholders and destinations with a high level of reliance on Thomas Cook clients is significant. When a company with a 178-year history collapses, without any convincing explanation from its senior management, the company's insolvency affects the reputation of the entire travel industry.

The Thomas Cook collapse is one of a series of high-profile business insolvencies in recent years. The Covid-19 pandemic and the subsequent demise of tourism businesses have accelerated the level of tourism business vulnerability. In August 2020, the collapse of STA Travel was one of many tourism businesses where operations were restricted by the regulatory limitations imposed on international travel businesses by Covid-19. Chapter 3 addresses the significant challenges to tourism-related businesses, which hitherto operated a business model based on high turnover and low yield. This vulnerability of tourism's 'default business model' was exposed due to the severe decline of travel bookings in all tourism industry sectors during the 2020 pandemic crisis.

Conversely, management errors or equipment failure, if handled sensitively and effectively, can actually enhance the reputation of a tourism destination or tourism business. The Fijian tourism industry's effective management of tourists affected by Cyclone Winston in 2016 actually enhanced Fiji's reputation as a destination, highly committed to the duty of care towards visitors during a natural disaster.

This chapter examines crises arising from substandard service, overpricing, inadequate responses to customer complaints, industrial relations problems and equipment or infrastructure failure. In many cases, the effective management of these issues can actually enhance the reputation of a destination or a tourism enterprise. Poor human resource management issues have the potential to create a corporate culture in which dissatisfied employees and poor people management practices can compromise the performance of enterprises in delivering optimal customer service.

Chapter 12 – Theme 9: Environmental Crises and Recovery

Climate change, alterations in weather patterns, pollution and human generated environmental degradation can singularly, or combined, have a significant impact on tourism destinations. Coral bleaching, rising sea levels, prematurely melting snowfields, and shifts in storm directions can undermine the environmental desirability of a destination. This chapter analyses how destinations can adapt their seasonality, product mix and marketing to environmental challenges.

A Positive Long-Term Prognosis for Global Tourism Resilience ... But

Despite many events which undermined the marketability and appeal of tourism throughout the world, the global tourism industry proved to be remarkably resilient until the beginning of 2020. In January 2020, the UNWTO tourism barometer indicated that global tourism movements in 2019 grew by six per cent over the previous year, or 50 per cent above the ten-year growth trend of four per cent per annum (UNWTO, 2020). Tourism flows to and from some destinations are significantly affected by crises. However, global tourism had demonstrated remarkable resilience to recover very quickly from actual and perceptual crises. The driving forces of tourism growth, including the growth of the middle class throughout the developing world and the increased affordability of travel, especially in the transport sector, advanced tourism growth far more than those factors which actually or potentially limit that growth.

The world's two most populous nations, China and India, grew far more rapidly as tourism generating countries than the most optimistic analysts could have ever imagined. By 2019, China was the world's leading generator of international tourism with

145 million Chinese travelling internationally, representing almost ten per cent of global outbound tourism. Yet the potential for growth is still significant as Chinese outbound tourist numbers in 2019 represented approximately ten per cent of China's population.

The outbreak of Covid-19 during 2020 exposed the underlying fragility of the tourism industry. Covid-19 has presented the greatest challenge to the resilience of tourism in living memory. New resilience theories now have to address a reality in which the viability of tourism operations has been severely undermined. Government regulations, border closures and social distance restrictions severely constrained the movement of people. Unlike recoveries from past crises, the recovery of tourism from Covid-19, subject to government policies and reliable medical treatment for the pandemic, may take years. The pre-Covid-19 tourism business model of high volume and low margins has revealed tourism enterprises as fragile and vulnerable. While the author remains confident that tourism will ultimately recover, the operation and management of post-Covid-19 tourism as 'the new normal' are likely to be different to what they were prior to the pandemic.

Managing tourism crises and expediting recovery from crises will increase in importance as the global tourism industry's significance within the global economy continues to grow. This book deals with strategies employed by destinations and tourism businesses to manage their recovery from a wide range of crises. The intention of this book is as a useful thematic guide for tourism scholars, researchers and practitioners alike. Crises, irrespective of cause, inflict pain, suffering and loss. We should never forget the human dimension of a crisis event. The effective management of risk and crises presents an opportunity for all businesses and enterprises to learn from and adapt to challenges and changing circumstances, which threaten the viability and reputation of all enterprises, especially tourism.

David Beirman, May 2021

REFERENCES

Avraham, E. and Ketter, E. (2016) *Tourism Marketing for Developing Countries*. Palgrave Macmillan: Basingstoke, UK. pp. 1–17.

Becken, S. and Carmigani, F. (2016) 'Syria Always Beautiful – can tourism be a source for peace?', *The Conversation*, 26 September 2016. www.theconversation.com. Accessed 30 November 2020.

Beirman, D. (2003) *Restoring Tourism Destinations in Crisis – A Strategic Marketing Approach*. Allen & Unwin: Crows Nest (Sydney), Australia.

Beirman, D. (2009) 'Crisis and post-crisis tourism destination marketing', in Pforr, C. and Hosie, P. (eds), *Crisis Management in the Tourism Industry: Beating the Odds*. Routledge: Farnham, UK. pp. 207–23.

Beirman, D. (2018) 'Thailand's approach to destination resilience: An historical perspective of tourism resilience from 2002 to 2018', *Tourism Review International*, 22: 277–92.

Beirman, D. and Walters, G. (2017) 'Marketing in the Middle East in times of political instability – The case of Jordan', in Almuhrzi, H., Alriyami, H. and Scott, N. (eds), *Tourism in the Arab World – An Industry Perspective*. Channel View Publications: Bristol, UK. pp. 175–87.

Gurtner, Y. (2007) 'Phuket: Tsunami and tourism – a preliminary investigation', in Chon, K., Laws, E. and Prideaux, B. (eds), *Crisis Management in Tourism*. Cabi Publishing: Wallingford, UK. pp. 217–33.

Henderson, J. (2007) *Tourism Crises: Causes, Consequence and Management*. Elsevier: Burlington, MA.

Hollway, S. (2007) *The Ten Commandments of Tourism Recovery*. Unpublished keynote speech delivered by Sandy Hollway to the APEC Tourism Forum. Surfers Paradise, Queensland, Australia on 15 May 2007. (The author was present and has used the material in teaching.)

Tarlow, P. (2014a) *Tourism Security*. Elsevier: Amsterdam.

Tarlow, P. (2014b) 'Tourism oriented policing in the tourism industry', *International Journal of Event Management Research*, 8(1): 1–24.

United Nations World Tourism Organization (UNWTO) (2020) www.unwto.org. Accessed 30 November 2020.

Wearing, S., Beirman, D. and Grabowski, S. (2020) 'Engaging volunteer tourism in post-disaster recovery in Nepal', *Annals of Tourism Research*, 80: 102802.

World Health Organization (WHO) (2020) www.who.int. Accessed 2 May 2021.

Worldometers (2020) *Covid-19 Coronavirus Pandemic*. www.worldometers.info/coronavirus. Accessed 2 May 2021.

World Travel and Tourism Council (WTTC) (2020) www.wttc.org. Accessed 30 November 2020.

PART 1

Establishing the Theoretical and Sectoral Context for Tourism Risk, Crisis and Recovery with focus on Covid-19 Tourism's Most Challenging Crisis

1

A CONTEXTUAL APPROACH TO TOURISM RISK, CRISIS, RECOVERY AND RESILIENCE

INTRODUCTION

This chapter focuses on defining the key terms that appear throughout this book. Since the mid-1990s, tourism risk, crisis and recovery have emerged as growing fields of scholarly research and study in tourism. The tourism and hospitality industries have also devoted increasing attention to the many external and internal threats to the viability and reputation of tourism destinations and enterprises.

As discussed in Chapter 2, some sectors of the tourism and hospitality industries have adopted a more proactive approach to dealing with risk and crisis than other sectors. Airlines and ocean cruising companies have long been aware that there are catastrophic risks associated with these forms of transport. Consequently, as sectors, they established governance protocols and policies to minimise exposure to and manage their key risks. By contrast, other sectors of the tourism and hospitality industries, notably travel retailers, destination marketing/management organisations and the accommodation providers, have tended to be reactive rather than proactive in responding to risks which threaten the viability of their businesses.

Until relatively recently, tourism scholars focused most of their attention on assessing the positives and negatives of tourism from a sustainability perspective and analysing the rise of mass tourism and marketing issues. Although risk and crisis were background issues for tourism scholars during much of the 1980s and 1990s, the impact of the 9/11 terrorist attack of September 2001, the Bali bombing of 2002 and SARS in 2003 led to a surge of scholarly and broader travel industry interest in the fields of tourism risk, crisis and recovery. This level of interest has coincided with the era of the internet and the widespread and instant dissemination of global news reporting.

DEFINING TOURISM RISK, CRISIS, RECOVERY AND RESILIENCE

A full consideration of post-crisis recovery and destination, organisational or enterprise resilience requires a clear understanding of the crises from which destinations and enterprises have recovered. The word *crisis* is frequently devalued in political, media, social or commercial discourse to refer to any set of untoward circumstances that may affect destination or enterprise performance in the achievement of optimal key performance indicators. Frequently, tourism destinations or businesses claim that they are experiencing a crisis because their visitor numbers were down five per cent or their profits fell by ten per cent below those of the previous year.

While such circumstances may legitimately constitute bad news and are indicative of a management or marketing problem, they don't necessarily constitute a crisis in which the viability, marketability or reputation of the destination or enterprise is either *severely* undermined or facing existential threat. The concept of *severity* is central to defining a crisis. Destinations and tourism enterprises experience cycles of growth and decline that reflect the external marketing or economic environment, the quality of strategic or operational management or the state of the competitive environment (Ritchie, 2009).

In setting a clear context for this book, a tourism crisis is defined as: 'An event or a set of circumstances arising from external causes or internal management failure which *severely* compromises the viability, reputation or marketability of a tourist destination or enterprise and requires significant rehabilitation and confidence building to restore viability, reputation or marketability'. This definition is based on an amalgam of definitions proposed by a range of distinguished academics, most notably, the late Bill Faulkner (Beirman, 2003; Ritchie, 2009; Wilks et al., 2006; World Bank, 2014).

The actual or perceived severity of an event or set of circumstances, and the degree of its impact on the destination or enterprise, represent core contextual determinants of a tourism crisis. Ritchie (2009) tracked the history of key definitions of crisis in the tourism academic literature from 1972 to 2005. Common threats that arose from those definitions were the concept of severity, surprise, negative impacts on reputation, security, public perception and a threat to the fundamental viability and marketability of a destination or an enterprise.

Although much of the academic literature on tourism crises tends to focus on crises arising from external threats such as natural disasters, war, terrorism, political instability, crime, pandemics and economic slumps (all of which are examined in succeeding chapters of this book), many crises, especially those affecting individual enterprises, arise from management and operational failure.

In the tourism risk management literature, a crisis (especially ones arising from management failure) frequently occurs when a destination or enterprise has not effectively factored in or planned for the treatment of those potential risks that could compromise the viability and the reputation of the enterprise. On 11 April 2017, United Airlines attracted a vast amount of public and media odium when it forcibly removed a paying passenger due to a practice (common throughout the commercial aviation sector) of overbooking (Benoit, 2018).

Overbooking, commonly practised by airlines, is predicated on the statistical likelihood that a number of booked and confirmed passengers tend to no-show on flights. In this case, there was an absence of no-shows. United's offer of financial compensation for passengers willing to voluntarily disembark was rejected by all passengers. United's unusually violent 're-accommodation' of this passenger, accompanied by the absence of a sincere apology, went viral on social media, losing it billions of dollars in lost business and reduced share value. The reputational damage to United was exacerbated by media accusations of racism against it when the expelled passenger was thought to be Chinese. He was actually a Vietnamese born American citizen but the incident generated over 100 million hits on the Chinese social media site WeChat in a single day. In the context of sensitive and strained political relations between the USA and China, since the outset of Donald Trump's presidency from January 2017, the incident had an unexpected and disturbing political and diplomatic dimension.

Had United Airlines developed a contingency plan for dealing with passengers who refuse voluntary disembarkation from a paid and confirmed flight, the ugly scenes of this passenger forcibly ejected from a United Airlines flight are unlikely to have occurred. As overbooking is common practice and entails a foreseeable risk to all airlines, treating the consequences of that risk should be a standard contingency management policy. Although the likelihood of such an incident was low, the consequences of mismanagement for United Airlines were high.

The translation from the Mandarin word for crisis means 'danger and opportunity'.

危机

Figure 1.1 The term 'crisis' in Mandarin. Mandarin lettering courtesy of Bert Van Walbeek 2020

This suggests that rather than treating crisis solely as a negative concept, there is the potential of a positive outcome, which arises either through the recovery process or through the introduction of enhanced risk management approaches to prevent, mitigate or circumvent reoccurrence of an impending crisis arising from similar causes. According to Hampton (2014), the core principle of enterprise risk management is employing risk management as a tool for continuous improvement of the enterprise.

A central theme of modern post-crisis recovery is the concept of 'building back better' (Hollway, 2007). Sandy Hollway is an Australian risk management professional who was appointed as chief organiser of the 2000 Sydney Olympic Games. Following the 2000 Olympics, the Australian government appointed Hollway in 2003 to lead the recovery of Australia's capital city, Canberra, after devastating bushfires destroyed hundreds of homes in Canberra's suburbs. Three years later, in 2006, the Australian government approached Hollway to manage the recovery of a number of towns in far North Queensland that had sustained extensive infrastructure damage from Cyclone Larry, which also destroyed 90 per cent of Australia's banana crop. Both of these recovery programmes included a significant tourism dimension. In both cases, the intent of the recovery programme was not merely to restore the destinations to the way they were prior to the disaster, but to enhance building codes and protective risk management practices in order to reduce the likelihood and impact of a similar future event having a similar impact.

The concept of building back better applies well beyond re-establishing infrastructure and management performance and the calibration of risk management procedures to a higher standard than pre-crisis levels, or even planning infrastructure to reduce its vulnerability to natural disasters or security threats. Building back better, in a tourism context, also includes the marketability and the market performance of a destination or enterprise exceeding pre-crisis levels. Resilience and growth are the points at which the concept of risk and opportunity meet. Many of the examples cited in this book refer to destinations and businesses that have endured a crisis, and through the lessons learned from both managing and overcoming the crisis have moved from stagnation to growth.

In a speech Hollway delivered to the Asia Pacific Economic Cooperation (APEC) Tourism Working Group, held in Surfers Paradise, Australia in May 2007, he outlined what he described as the 'ten commandments of destination recovery' from a crisis event (Beirman, 2016). In summary, they are:

1. Collaboration between the private sector and the public sector.
2. Dedicated machinery and resources devoted to recovery within agreed time limits.
3. In the case of destination recovery, all tiers of government should be involved.
4. Rapid response, especially in dealing with traditional and electronic (including social) media.
5. Engagement and support for the affected community.
6. Public communications, which are centralised, honest and informative.
7. Proper process with planned outcomes.

8. Importance should focus on people ahead of budgets.
9. Strategic and contingency planning to get ahead of the curve.
10. A focus on building back better.

There is a significant correlation between the concepts outlined by Hollway as a practitioner and the extensive body of academic literature, which has addressed recovery and destination resilience. Many academics (Calgaro et al., 2014; Cochrane, 2010; Dwyer et al., 2009; Jamal and Stronza, 2009) have focused on the importance of collaboration in the implementation of destination recovery. In the context of social sustainability, the recovery of tourism destinations that experienced a natural disaster, terrorism, pandemic, a crime wave or political unrest need to be guided by the readiness of the affected community to accept a resumption of tourism and the terms of that acceptance. Figure 1.2 illustrates the model of tourism resilience developed by Cochrane (2010). Central to this model is an understanding that an effective response to a crisis involves flexibility, adaptability and stakeholder cohesion. In fact, virtually every successful post-crisis recovery campaign in global tourism has incorporated these core elements. The elements in this model are especially relevant to the initial response to a crisis.

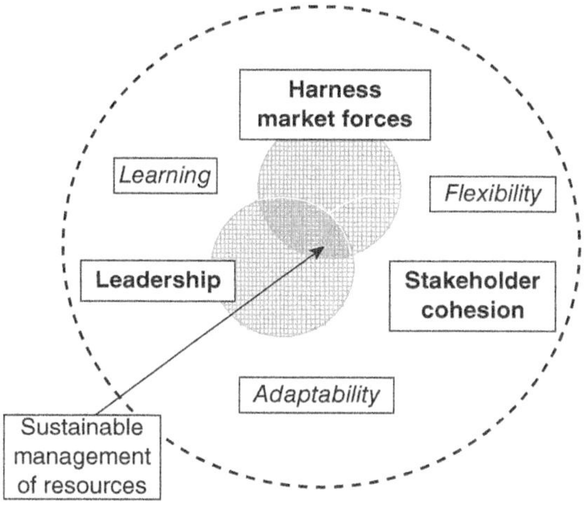

Figure 1.2 The sphere of tourism resilience

Source: Cochrane, 'The Sphere of Tourism Resilience', *Tourism Recreation Research* (2010). Reprinted by permission of Taylor & Francis Ltd.

Crisis communications is a significant element in the academic literature of tourism recovery (Avraham and Ketter, 2008 and 2016). Issues including timing, transparency, verifiability, consistency and relevance are important in setting and establishing the legitimacy of a destination's or an enterprise's messaging to its stakeholders and customers during a crisis and throughout its recovery phase. Today, crisis communications requires that enterprises should effectively employ multiple platforms to communicate messages, including traditional media, social media, websites and other forms of online communication.

In 2003 the Pacific Asia Travel Association (PATA) developed a strategic approach to crisis communications in response to the severe impact the fear of severe acute respiratory syndrome (SARS) imposed on tourism to South-East Asian countries. As PATA's then CEO Peter de Jong described it, Project Phoenix, 'was a reputation management strategy in which PATA, in association with its wide range of travel industry members and stakeholders, worked cooperatively to restore confidence in and reputation of SE Asian destinations' (Yates, 2006). The media, especially the BBC and CNN International cooperated closely with PATA to promote a message that travellers to South-East Asia had little to fear from SARS and were welcome to return. A core intention of Project Phoenix was to counter and challenge many of the negative perceptions surrounding SARS in South-East Asia with facts and context. Having two major global TV networks (CNN and the BBC) as allies in this cause presented a challenge to media outlets, including CNN and the BBC, which played a major role in creating some of the fear about visitation to South-East Asia resulting from sensational coverage of the disease and its impact. Project Phoenix was highly successful in stimulating the rapid recovery of many SARS affected destinations (most of which were in the Asia Pacific region). Consequently, it became a widely used template for many tourism destinations and enterprises engaged in recovery from crises.

The challenge of Covid-19 during 2020 has led tourism organisations all over the world to revise crisis communications strategies. The World Travel and Tourism Council (WTTC) released a series of Covid-19 tourism protocols in July 2020. PATA (2020) released a crisis communications toolkit in August 2020, which placed a great deal of extra emphasis on social media and the 24/7/365 news cycle. PATA is not alone in addressing this critical issue. Most national and global tourism associations have reassessed their media engagement in response to a crisis, which emerged as an existential threat to businesses in all sectors of the global tourism industry.

RESILIENCE AND TOURISM

Closely related to the concept of recovery is that of *resilience*, which applies to tourism enterprises, sectors and destinations. In the context of tourism, resilience involves the ability of an individual enterprise, a sector or destination to recover rapidly from negative events and circumstances to the extent that reputation, viability and marketability compromised or lost during the crisis are rapidly restored (Hall et al., 2018). This goes beyond addressing the problem/s that may have led to a crisis; also instilling reputation, trust and confidence among consumers and stakeholders of the destination or enterprise's viability.

The destination sustainability model in Figure 1.3 developed by Calgaro et al. (2014) served as a comprehensive strategic approach to destination resilience with an emphasis on sustainability. It accounted for the multiplicity of crisis causalities and set the spatial, temporal, geographic, infrastructure and social boundaries of a crisis. It also included the impact of a crisis and the appropriate operational and communication responses, plus the governance of those responses in managing the recovery from a tourism-related crisis event for a destination or a tourism enterprise.

Shock and stressor characteristics	**Exposure**	**Sensitivity**	**System adaptiveness**
Typology (e.g., natural disaster or social disruption)	**Population-size, characteristics and distribution**	**Tourism-specific sensitivities**	**Impact and coping responses, including emergency responses and aid provision**
Magnitude	**Biophysical environment**	Includes seasonality, source markets, destination history, market positioning and destination image	**Adjustments and adaptations**
Frequency	Natural terrain	**Economic issues**	Technical
Duration	Climate	Includes economic dependence on tourism	Behavioural
	Ecosystem characteristics	**Human and social**	Business management
	Usage of natural resources	Includes enterprise and employee skills, labour capacity, risk management skills	Policy
	Built environment	**Physical and environmental**	Research and education
	Settlement location	Access to natural resources, extent and quality of tourism infrastructure and transport	
	Development pattern type and location	**Governance processes**	
		Government responsibility	
		Laws, policies and rights	
		Cultural governance	
		Tourism business networks	
		Degree of political and social stability	

Figure 1.3 The destination sustainability framework

Adapted by the author from Calgaro et al., 2014, with kind permission of Dr Emma Calgaro, University of Sydney

In essence, the model addressed two of the key pillars of contemporary tourism development: risk management and sustainability. Neither can be mutually exclusive. The prevailing attitude among tourism scholars, practitioners and policy makers is that post-crisis recovery and destination resilience are aligned to sustainable tourism principles. Post-crisis tourism recovery programmes which ignore the willingness of an affected community to support a resumption of tourism can potentially be socially problematic.

In the six months following the 9/11 attack of September 2001, which fundamentally altered the risk and security environment for the global airline and airport sectors, consumer confidence in the safety of airlines and airports, slumped. The approach of the International Civil Aviation Organisation (ICAO), the International Air Transport Association (IATA) and the Airport Council International (ACI), which between them represented all global airlines and airports, was to collaborate in developing a comprehensive range of post-9/11 security policies and protocols.

Upgraded safety measures and protocols were developed. The implementation of these protocols was designed to restore confidence among all relevant stakeholders, especially prospective airline passengers, that the airline and airport sectors had comprehensively addressed safety and security failures arising from 9/11. The global airline industry operates under global governance through the ICAO, which is a UN body empowered to impose base standards for safety on all carriers in the world. However, the development of upgraded security protocols was a cooperative venture in which airlines and airports reached a consensus. The fact that the number of airline passengers began to increase after March 2002 was evidence of a *recovery*, but the fact that travellers felt the confidence to fly on commercial passenger services was evidence of the sector's *resilience*. Security consideration limited the details airlines and airports could publicly disseminate about many of the new security measures introduced. However, in order to boost consumer confidence, all stakeholders were reassured that enhanced security measures would apply to all airlines and airports globally.

In recent years, the ICAO's proactive approach to airline safety risks averted several potentially catastrophic events affecting airlines. Volcanic ash clouds represent a major threat to the operation of jet airlines through their ability to render a jet engine inoperable. In 2009, the ICAO (ICAO, 2009) introduced a series of global protocols that required airlines to either avoid areas with specific densities of volcanic ash or cease operation in that area. During one week in April 2010, much of Western Europe closed air traffic due to the high density of volcanic ash particles in the atmosphere originating from the Eyjafjallajokull volcanic eruption in Iceland. In November 2017, air traffic to and from Bali was suspended due to the eruption of Bali's Mount Agung. While both incidents severely disrupted airline flight schedules and associated tourism transport, no deaths or injuries occurred.

COMMUNICATING RECOVERY AND RESILIENCE

Some destination management organisations, in their anxiety to regain the economic benefits of tourism lost during a crisis event, ignore the safety concerns of tourists in returning to a destination in which a political crisis remains unresolved. The 'Arab Spring' political demonstrations, in Cairo during February 2011, ultimately precipitated the overthrow of former Egyptian President Hosni Mubarak and was accompanied by a massive exodus of international tourists from Egypt (Avraham, 2015).

In the weeks following Mubarak's overthrow, the Egyptian Tourism Authority (ETA) launched a campaign to urge tourists to return to a 'new democratic Egypt'. Regrettably, the ETA's call was premature. Mubarak's fall from power in February 2011 did not transform Egypt into a model democracy. Contrary to the ETA's enthusiastic and premature claims, the ongoing political instability prevented a genuine transformation in Egypt, undermining the ETA's ability and credibility to market Egypt as a truly safe tourism destination. Conversely, it ushered in a period of ongoing political instability and security threats to tourists between 2011 and 2015. This included the brief rise of Mohamed Morsi's Muslim Brotherhood government between 2012 and 2014, and its subsequent overthrow by a military coup led by Egypt's current President General Fattah el-Sisi in 2014. By 2019, Al Sisi's government was subject to growing opposition.

As discussed in Chapter 5, Egypt's political instability was a breeding ground for multiple acts of jihadist terrorism between 2011 and 2019, some of which targeted tourists. The ETA tended to either ignore these incidents on its website or give vague assurances that security measures to protect tourists were in place (Avraham and Ketter, 2008).

As a corollary, the ETA's failure to inform its many stakeholders of safety threats and the facts on the ground in Egypt since 2011 lost it considerable credibility among a range of Egyptian and international stakeholders. The ETA failed in its duty of care to be a reliable and trustworthy source of destination information about Egypt. For a destination management organisation (DMO), loss of trust and credibility damages its reputation among travel industry and consumer stakeholders, one of their most precious intangible assets. Simultaneously, the reputation of the destination that the DMO represented was also undermined.

The anti-Beijing political unrest in Hong Kong, which began in June 2019 and escalated for some months into late 2019 and throughout 2020, transformed Hong Kong's reputation and perception as one of the world's most popular and welcoming stopover destinations to that of a risky and undesirable tourist and convention destination.

Hong Kong's perceptual problems were exacerbated by the self-censorship undertaken by the Hong Kong Tourism Board (HKTB) in its non-referral to the protests in Hong Kong, some of which involved occupying the international airport (HKTB, 2021). The HKTB has done all it can to adopt the three wise monkies approach to reputational risk: see no evil, hear no evil and speak no evil. The fear of a negative response from Beijing appears to override the HKTB's duty of care to inform travellers and the travel industry about the disruption to tourism and tourism infrastructure resulting from anti-regime protest action in Hong Kong.

The author participated in a number of post-crisis tourism recovery programmes, including the post-2015 earthquake recovery in Nepal (Beirman et al., 2018; Wearing et al., 2020). In each case, the host community's preparedness to accept tourist visitation varied from those destination communities who may wish to have an extended moratorium on tourism during a reconstruction and grieving phase, to those communities who want to welcome tourists immediately. The timing and scale of a tourism recovery programme should respect and account for community attitudes. This is a central axiom of socially sustainable tourism.

A feature of Nepal's tourism recovery after the April 2015 earthquake was the significant emphasis on volunteer tourism in which these tourists participated in the post-earthquake infrastructure recovery programmes in conjunction with the affected communities. In Nepal, volunteer tourism was a major contributing factor to its relatively rapid tourism recovery. In 2015, the year of the earthquake, 560,000 tourists visited Nepal, representing a 35 per cent decline in international tourism compared to the previous year. By the end of 2018, international tourism numbers reached 1.25 million for the year, which is an all-time record (Government of Nepal, 2020). Nepal's recovery was largely driven by a marketing strategy devised by PATA and PATA's Nepal chapter (PATA, 2015; Wearing et al., 2020), which focused on extensive employment of social media to communicate the recovery process of the destination's tourism infrastructure and its readiness to welcome and accommodate tourists, and also publicise positive experiences of visitors. Celebrity endorsements, including the visit to Nepal in March 2016 by Britain's Prince Harry and the promotion of volunteer tourism, enhanced the popularity of volunteer tourism as an integral element of Nepal's tourism experience (Wearing et al., 2020.)

However, during May and June 2019, Nepal's newly restored image as a safe tourism destination was compromised due to the lack of governance applying to mountaineering tourism. During that period, 11 out of 891 climbers who attempted to summit Everest died. The media was full of disturbing images of hundreds of climbers queueing to summit Everest. This resulted in long waits among climbing parties, exposing them to severe health and climatic dangers. These risks included lack of oxygen and prolonged exposure to the elements. The images of avaricious operators profiting from climbers' aspirations in a very risky environment for climbers, operator employees, including guides and porters, inflicted significant reputational damage on Nepal as a tourism destination.

JAPAN'S RAPID TOURISM RECOVERY FROM THE 2011 EARTHQUAKE AND TSUNAMI

Although Okinawa was Japan's most distant region from Sendai, the epicentre of the devastating 2011 earthquake and tsunami (2.5-hour flight time south from Tokyo) and Japan's least affected region, Okinawa's prefectural government commissioned a massive re-evaluation of preventative measures for earthquakes, tsunamis and typhoons. These included enhanced building standards and siting regulations, evacuation and crowd assembly plans applying to public buildings, shopping centres, hotels, sport stadia, event venues and entertainment precincts. In his analysis of Okinawa's approach, Masato Takamatsu (2014), CEO of Japan Marketing, used Okinawa's proactive approach to risk management as a vehicle to incorporate Okinawa's commitment to safety as a core destination marketing asset.

Japan's location on the Ring of Fire results in a high level of vulnerability to natural disasters, especially earthquakes, volcanic eruption, tsunamis, sea surges and typhoons. A proactive approach to risk management led Japan to become a global leader, in developing proactive risk management policies and practices.

Japan's impressive 500 per cent growth of international inbound tourism arrivals from 2011 (6.2 million tourists) to 2019 (31.9 million tourists) illustrated by Figure 1.4 has been largely predicated on emphasising Japan's commitment to tourism safety as a core destination attribute. This growth of tourism to Japan was well in advance of Tokyo's intended hosting of the 2020 summer Olympics.

In common with all international tourism destinations, the Japanese government's forecast for tourism growth during 2020 experienced a reverse due to the Covid-19 pandemic. The planned surge of tourism during the scheduled 2020 Olympics was derailed.

THAILAND: A GLOBAL ROLE MODEL OF TOURISM DESTINATION RESILIENCE

Until the onset of the Covid-19 pandemic, Thailand was highly successful in adapting to and rapidly recovering from a wide variety of negative events. Thailand has been widely acknowledged as one of the world's most resilient tourism destinations. It has responded to multiple challenges, which would have resulted in long-term

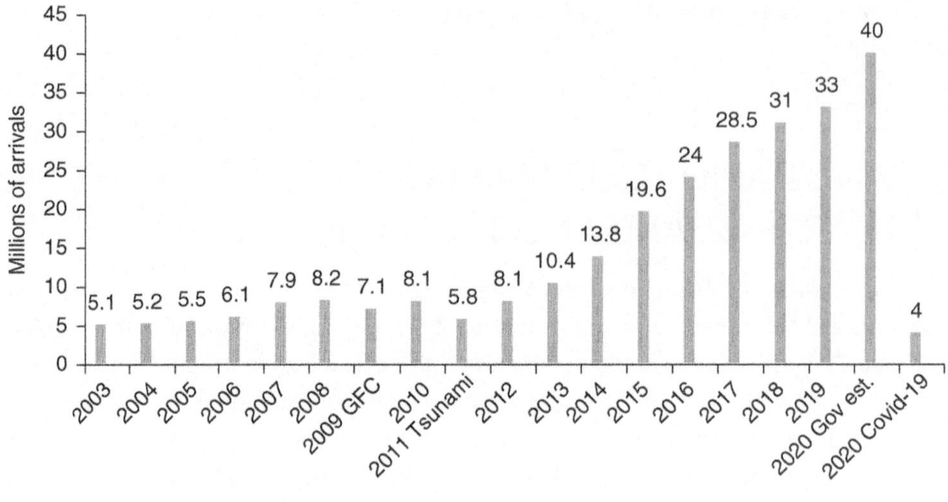

Figure 1.4 International tourism arrivals to Japan, 2003–2020 (millions)

Source: Japan National Tourism Organisation, 2020. Graph by the author

setbacks for tourism in most other destinations. Thailand is not alone among destinations contending with more than one reputational challenge at once. During 2016, Brazil responded to concerns about an outbreak of Zika and media attention on its crime rate, while hosting the 2016 Rio Olympics. Figure 1.5 demonstrates that between 2003 and 2016 Thailand achieved a high level of annual growth of international tourism arrivals in the face of multiple significant challenges, ranging from natural disasters and terrorism to pandemics, political unrest and the global financial crisis. Figure 1.5 shows tourism growth to Thailand between 2004 and 2019. The onset of Covid-19 in 2020 led to a steep decline in international tourism arrivals in line with a global trend resulting from the closure of most international borders.

The Thai tourism industry had demonstrated a clear understanding of the importance of crisis management since 1998 when it responded to a significant downturn in tourism resulting from the Asian Financial Crisis during that year. The Tourism Authority of Thailand has demonstrated a propensity to embrace crisis management expertise from both local and international sources. The author witnessed this in January 2005 when the UNWTO chose Thailand as the venue to develop an emergency recovery campaign for tourism to all countries affected by the Indian Ocean tsunami of 26 December 2004.

In 2003, Thailand's tourism arrivals declined due to the SARS outbreak, prevalent throughout South-East Asia. Despite the fact that Thailand recorded only 12 cases of SARS during 2003, it experienced a 1.5 million drop in the number of tourists compared to 2002. This was a classic example of a perceptual crisis.

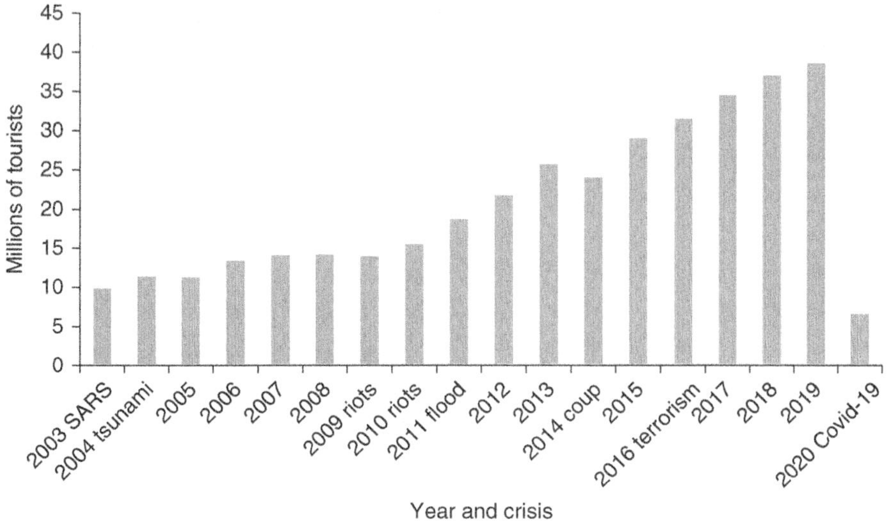

Figure 1.5 Thailand's tourism growth and the Covid-19 decline, 2003–2020

Source: Tourism Authority of Thailand, 2020. Graph by the author

Despite the significant loss of life (especially in Phuket and Thailand's Andaman Sea coast) from the Indian Ocean tsunami of December 2004, the tourism arrival figures of 2005 registered a very small decline. Episodes of political instability, some erupting in violent clashes between supporters and opponents of the government, were prevalent in Thailand between 2008 and 2010. Yet, they had a modest impact on the growth of international tourism to Thailand. This period also coincided with the global financial crisis, which had minimal impact on tourism arrivals to Thailand. In the case of a financial downturn, Thailand's actively promoted, low price value proposition for tourism services actually provided an incentive for visitation from budget conscious tourists. The massive floods, which inundated many parts of central Thailand during November 2011, had little impact on tourism arrivals largely due to the Tourism Authority of Thailand's effective and accurate information campaign, run from its website and communicated via social media. An ongoing civil war between Muslim separatists and Thai government forces on Thailand's southern frontier region with Malaysia had minimal impact on tourism, partially because the Thai Tourism Authority was able to assure tourists that these problems were distant from popular regions of visitation.

The only recent year since 2003 (apart from 2020) in which there was a notable downturn in tourism to Thailand was during 2014. Thailand experienced intense political instability climaxed by a military coup that installed a military government, which continues to rule Thailand today. Acts of terrorism in 2015–2016 and the death of Thailand's revered monarch King Bhumibol in October 2016, ending a reign of 70 years, had minimal impact on the growth of tourism. The death of the Thai monarch led to a month of national mourning and the cancellation of many events, which traditionally attract many international tourists. Yet, despite some media predictions of Thailand experiencing a slump in tourism during the final quarter of 2016, the year marked significant growth in tourism arrivals.

Thailand's tourism resilience and its ability to rapidly recover from negative events owes much to the marketing agility of the Tourism Authority of Thailand to effectively and rapidly acknowledge, address and respond to negative events. Successive Thai governments, irrespective of political persuasion have long recognised that tourism is one of Thailand's most important sources of income and a major generator of employment. Tourism is widely seen as a significant industry (representing 20 per cent of Thailand's GDP) that is worth protecting, promoting and preserving (WTTC, 2020).

The Tourism Authority of Thailand has demonstrated a high level of commitment to adapt its tourism marketing to a range of challenging situations. The author has witnessed Thailand's effective recovery from the December 2004 Indian Ocean tsunami, its recovery following the political upheavals of 2008–2010, the military coup of 2014 and the death of King Bhumibol in October 2016. Without detracting from the considerable skill and determination of the Tourism Authority of Thailand's management and staff, its proven ability to recover quickly from potential crises has benefited from its ready access to the management skills of PATA, headquartered in Bangkok since 1998. Since its move to Bangkok, PATA has been a leading transnational tourism association in developing tourism risk and crisis management protocols, training programmes and effectively disseminating those skills to its government and private sector members in over 70 countries (PATA, 2020).

A number of key elements have contributed to the Tourism Authority of Thailand's success in recovering from a range of negative events and building the resilience of 'destination Thailand' (Beirman, 2018):

1. The Tourism Authority of Thailand (TAT) consistently responds quickly to any negative events and communicates its response to a comprehensive range of stakeholders.
2. In the event of a natural disaster or political unrest, the TAT rapidly communicates the status of tourism access to key attractions, the operability and availability of transport and accommodation, and the status of events in real time and with a commitment to accuracy.
3. The TAT has a consistent commitment to transparency and demonstrates preparedness to facilitate media and travel industry opinion leaders from key source markets to come to visit Thailand as its guest to see for themselves.
4. The TAT makes extensive use of travellers' opinions disseminated on its social media platforms.

5. The Tourism Authority of Thailand demonstrates agility in adapting its marketing programmes in response to a changed situation. In October 2016, during the intense national mourning period following the death of King Bhumibol, the TAT asked visitors to demonstrate appropriate respect and decorum during their visit. Although many of the more noisy and celebratory events were cancelled in Thailand during the mourning period, the TAT focused its marketing on cultural tourism, which attracted a hitherto untapped market.

While the Tourism Authority of Thailand has adhered to and in most cases exceeded most key performance indicators of crisis and recovery management, it has shown a marked reluctance to engage in proactive communication of tourism security risks. The justification for this reluctance was predicated by concern that Thailand may lose face, if its tourism authority openly canvassed responses to safety threats in Thailand. Ironically, in 2013 the author and British tourism academic Alexandros Paraskevas were invited to Bangkok by the Tourism Authority of Thailand to speak on the issue of proactive tourism safety advice at a Thai tourism leadership conference in Bangkok jointly organised by the Tourism Authority of Thailand, PATA and global tourism industry networking group SKAL (Citonot, 2013). The fact that these safety threats are likely to apply to many other destinations is not considered. While Thailand responds effectively when a problematic issue arises, the Tourism Authority of Thailand remains reluctant to unilaterally adopt the proactive approach to tourism safety, which South Africa employed with considerable success to address tourism concerns about crime in 2000 (Beirman, 2003).

In May 2017, a meeting was held in Bali of the Asia Pacific Economic Cooperation (APEC) Counter Terrorism Working Group, which focused on counter-terrorism approaches to protect tourism. The 19 governments represented at the conference, including Thailand, agreed that their national tourism authority would provide incoming tourists with a range of pre-arrival security and safety guidelines. It is appropriate to note that there was no specified timetable for implementing these measures (APEC, 2017).

The important issue arising from examining Thailand is that a core element of its tourism resilience, in the face of a variety of negative events, involved accumulating a high level of credibility and trust among its many stakeholders. As a result, when the Tourism Authority of Thailand announces that Thailand is ready to welcome travellers, it is widely believed. Thailand is a sufficiently diverse destination to enable a downturn in tourism in one part of the country to be offset by tourism growth in another part. In the weeks following the December 2004 Indian Ocean tsunami, there was a significant decline in tourism to the Andaman Sea coast and most notably Phuket, which experienced severe damage from the tsunami. While tourism numbers to tsunami-affected parts of Thailand declined, tourism to unaffected parts of the country, notably Bangkok, inland Thailand and the Gulf of Thailand's coastal and island destinations, grew until Phuket could resume normal tourism activities.

RISK MANAGEMENT AND ITS RELATIONSHIP WITH TOURISM RECOVERY AND RESILIENCE

In common with crisis management, for many decades most tourism businesses treated risk management, with the honourable exception of the transport sector, as a peripheral management skill. The greatest change in the field of risk management in recent years was its transformation in many governmental and legal jurisdictions from a discretionary element of good corporate citizenship, evolving to a legally binding series of business practice requirements.

Risk management incorporates a range of concepts, including duty of care, responsible service of food and alcohol, occupational health and safety laws. Enterprises take responsibility for the safety and wellbeing of their customers, passengers, guests, visitors and staff while engaged in tours, visiting or staying at the premises or participating in the programmes of the travel company (Dunn, 2012).

In essence, risk management is about the calculation and prioritisation of the likelihood and consequences of potential threats to the safety, service, product, operational and financial success and customer satisfaction of a business.

Risk management (or contingency management) is predominantly the proactive dimension of risk, crisis and recovery management. A risk audit involves a prioritisation of threats to the viability of an enterprise and a summary of how these risks are to be treated. The basis on which risk is prioritised and treated reflects the likelihood of the risk arising and the possible consequence of the identified threat occurring. The International Standards Organisation (ISO) document ISO 31000–2018 and its predecessor ISO 31000–2009 sets out the contextual approach to risk management applicable to all enterprises. In essence, the 2009 version International Standards Organisation risk management flow commenced with the establishment of the risk context, a clear identification of the risks

To the enterprise, destination or organisation. This is followed by analysing risks, evaluating (and prioritising) risk, evaluation risks and then treating the identified risks.

In order to facilitate the flow, the ISO model recommends regular communication and consultation with stakeholder in determining risk management treatment. After a given risk or crisis event has passed, the final stage in the flow chart involves monitoring and evaluation of how the risk was managed. The following link will illustrate these points: https://www.iso.org/obp/ui/#iso:std:iso:31000:ed-2:v1:en.

While the ISO's core approach to risk management has changed little between 2009 and 2018, the ISO's 2018 revised risk management document placed far more emphasis on the importance of leadership, value creation and reputation enhancement than the 2009 document.

Each tourism enterprise or destination will necessarily have different levels of risk prioritisation and treatment and a varying range of stakeholders with whom they need to communicate and consult. As most tourism enterprises work with others towards the delivery of the customer's total tourism experience, collaboration in developing a risk management audit becomes important.

Whenever possible, enterprises seek to avoid owning direct responsibility for risk unless the alternatives are too expensive or unviable. The transfer of risk involves passing on the financial responsibility of a given risk to another individual or organisation. In many cases, this involves a financial contract with an insurance company accepting a given risk for a price (the premium). Premiums charged by insurance companies are based on a statistically calculated measurement of the likelihood and (financial) consequence of a given risk or set of risks (Insurance Council of Australia, 2019).

Contracts drawn up between tourism providers and their clients usually include a series of disclaimers of responsibility. However, written contracts, which deny the obligation of a service provider's responsibility for liability in breach of jurisdictional law, can be overruled in the course of litigation. It is vital for all tourism enterprises to ensure that their contracts of sale or service are written in accordance with the prevailing duty of care, occupational health and safety, legislation and associated obligations applying to the legal jurisdiction in which the contract applies (Cordato, 2020).

THE SECTORAL DIMENSION OF TOURISM RISK, CRISIS, RESILIENCE AND RECOVERY

The study and management of tourism risk, crisis and recovery includes an important sectoral dimension. One of the most challenging aspects of tourism, especially in the field of risk crisis and recovery, is the highly fragmented nature of the tourism industry, which operates under a number of broad sectors. Some elements of the tourism industry such as the airlines, airports and cruising are receptive to and accepting of global governance and common sets of protocols, applicable to risk and crisis management. By contrast, the accommodation sector involves a vast range of accommodation types, not readily able to operate under a common set of policies and protocols. Chapter 2 addresses the challenges applying to travel industry sectors in more detail. However, as this chapter is focusing on contextual approaches to risk and crisis, sectoral considerations represent an important context.

Every sector of the tourism industry has a role in risk, crisis and recovery management. These include:

- Transport which is divided in accordance with the mode of transport and associated port/terminal/airport facilities.
- Accommodation, which is divided into a significant range of subsets, which vary according to the size of the enterprise, the nature of accommodation they seek to provide and the demographics of the customers serviced.
- Attractions, which also contain a wide range of subsets and ownership types.
- Tour operators and tour wholesalers operating as inbound or outbound operators and in some cases both.
- Travel retailers/agents operating online, face to face or both. Included in this sector are travel management companies focusing on the corporate market.
- Tour guides.
- Destination management/marketing organisations.
- Hospitality and leisure services, including restaurants, theatres, cinemas, health centres, sporting venues, fitness centres, concert halls and opera houses.
- Meetings, incentives, conventions, event managers, organisers and venues.
- Transnational, national, state and local government tourism policy instrumentalities.
- Volunteer and special interest tourism services.
- Educational institutions and training providers.

A quick glance at the above sectors indicates that some sectors are more conducive to global risk crisis and recovery management protocols and governance than others. As previously discussed, the airline sector operates under global governance imposed by the International Civil Aviation Organisation, a United Nations (UN) instrumentality. Since the Wright brothers' first powered flight in 1903 there has always been a clear understanding of the possibility that an aircraft may crash. As commercial airlines have grown in size, sophistication and safety, the likelihood of an airline crash is very small but the consequences of an air crash can be catastrophic, especially for the world's largest civil aircraft the Airbus A380, which can carry over 500 passengers. However, global aviation has been the lead sector in tourism in the development and implementation of universal risk management practices, policies and protocols.

The cruise sector, in common with airlines, also treats risk and crisis management seriously. Modern cruise ship technology has greatly reduced the likelihood of a cruise ship sinking. Modern cruise ships have now reached a size in which the world's largest cruise vessels are travelling with over 6,000 passengers and almost 3,000 crew. The potential consequences of a cruise disaster are apocalyptic. Large numbers of people in a relatively confined space create a heightened risk of contagious disease. This risk manifested itself during numerous cruises in recent history. This risk assumed global prominence during the 2020 Covid-19 pandemic. Global cruising's risk management protocols were determined by the International Maritime Organisation (a UN instrumentality, since 1948). Global shipping has operated under a set of risk management protocols called SOLAS (Saving of Life at Sea 2020) first formulated in 1914. Due to global cruising's domination by a relatively small number of large companies, enforcement of standards is rigorous.

Global and national approaches to risk management operate widely in the transport sector. Although rail and road transport lack global approaches to risk management they tend to be strictly governed at the national or provincial/state level and sometimes at the transnational level, especially in the European Union. However, at the opposite end of the governability scale are the extreme difficulties in developing common risk management approaches to the exceedingly diverse accommodation sector.

The sheer diversity of the accommodation sector makes it almost impossible to set a global set of risk management standards or practices. Commercial accommodation literally covers the range from hiring out a sofa in a private apartment, selling camping space in a caravan park, a bunk in a backpacker's hostel, through to a bed and breakfast home stay or a suite at a 1,000 room five-star hotel or resort in any of the world's great city hotels or seaside resorts. In theory, the global hotel industry is represented by the International Hotels and Restaurant Association (IHRA) (IHRA, 2020), but this is essentially a networking and advocacy association for the hotel and restaurant industries. It has neither political power nor legitimacy to impose governance on its members. It also does not represent many providers of accommodation, including pubs, caravan parks, hostels, guest houses, private holiday rental housing, private inns and bed and breakfast establishments, religious hospice accommodation, serviced apartments and certainly not the more recent phenomenon of the shared economy's accommodation colossus, Air BnB.

Overwhelmingly, the global accommodation sector is dominated by private enterprises, most of them small. Most accommodation providers are subject to local, state or national governance relating to building standards, fire prevention, occupational health and safety, responsible service of food and alcohol, and workplace health and safety regulations. However, these regulations vary considerably between legal jurisdictions. There has been a great deal of recent attention given to security in hotels and other accommodation providers.

Accommodation providers face a functional and operational dilemma. All are part of the 'hospitality industry' (which also includes restaurants and places of entertainment) and this suggests that hoteliers should be open and welcoming to all guests whether they are staying in a hotel room, attending an event in the hotel, dining in a hotel restaurant or cafe or meeting friends in the hotel lobby. Conversely, hoteliers have a duty of care for the safety of their guest and governments frequently propose that those accommodation providers deemed as the most likely targets of crime or terrorism should screen their guests.

This hotel security dilemma came to a head on 1 October 2017 when Stephen Paddock was alleged to have opened fire from his guest suite on the 32nd floor of the Mandalay Bay Casino Resort in Las Vegas on an audience attending a country concert just outside the hotel grounds. In a ten-minute shooting spree, using over ten different rifles fitted with bump stocks, which enabled semi-automatic rifles to fire like a machine gun, Paddock killed 58 people and wounded 540. Although the massacre neither targeted the hotel nor its guests, many questions were asked by media and security specialists as to how a hotel guest could smuggle a veritable arsenal of weapons and ammunition into a hotel room (Beirman, 2018).

The Las Vegas shooting is among the deadliest attacks launched either in or from a hotel. Unfortunately, places of accommodation have been frequent venues for murderous attacks by terrorists and criminals. The 2008 terrorist attack on the Taj Mahal Palace Hotel in Mumbai (Rabassa et al., 2009) and recent suicide bombings and armed attacks targeting hotels in diverse locations, including Mali, Tunisia, Burkina Faso, Egypt, Jordan, Pakistan, Kenya, Sri Lanka and Jakarta, have raised questions about the need for screening of hotel guests and their baggage. The lack of any centralised governance applying to the accommodation sector makes it almost impossible to regulate from a sectoral perspective. However, many national, state and local government jurisdictions are now placing legislative pressure on hotels within their jurisdictions to upgrade security. While many of the world's largest accommodation chains, including the Intercontinental Hotel Group, Accor, Marriott and Hilton, do have detailed risk management policies and security protocols in place many other accommodation providers do not. Even after the Las Vegas shootings the Las Vegas tourist board and MGM Resorts International, the parent company of the Mandalay Bay Resort, was reluctant to call on its branded hotels to upgrade guest and baggage screening (MGM Resorts, 2020).

Concerns about the cost of baggage screening coupled with the belief that screening would conflict with the hotel's role as a provider of hospitality have dominated the reluctance of many of the world's accommodation enterprises to upgrade security measures applying to visitors or guests.

The issue of security screening is not a core issue for many accommodation providers. Privately owned bed and breakfast establishments, pubs, small hotels and motels and most of the properties operating under Air BnB are designed to accommodate a small number of guests and are unlikely to be subject to terrorist threats. However, the majority of terrorist attacks have targeted high-end hotels and resorts, which attract a large number of international guests.

The recovery of enterprises within a given sector and the rebuilding of reputation and trust after a crisis event are far more likely if an entire sector has introduced a systemic response to the crisis. In the six months following the 9/11 attack in September there was a significant slump in global civil aviation. However, airlines and airports worked globally to introduce a range of sector-wide security upgrades applied globally to all airlines and airports. The widespread dissemination of this process, if not the specific security measures, stimulated a resurgence in global aviation after mid-2002.

Some hotel chains targeted by acts of terrorism have made their stakeholders aware of the security upgrades they have introduced. This is far from a universal practice within the accommodation sector. In cases where this has occurred, the chains involved have sought to present these measures as part of their commitment to duty of care and as a PR and marketing benefit.

Destinations, which appreciate the value of building and maintaining a good reputation with their stakeholders, have incorporated attention to risk management as an integral part of the destination image. At the transnational level, destinations are strongly encouraged to achieve high standards in destination risk management. In 2011, the UN World Tourism Organisation released its Crisis Communications Toolkit (UNWTO, 2011). In 2011, the author and tourism risk management authority Dr Bert Van Walbeek co-authored the PATA risk, crisis and recovery management manual *Bounce Back* (Beirman and Van Walbeek, 2011). PATA and the WTTC developed updated risk management protocols and strategies (WTTC, 2020) in response to Covid-19, discussed in Chapter 3.

While the manuals and associated webinar teaching modules were used by DMOs and other tourism businesses, there is no international governance obligating the destination management sector to adhere to the advice given on risk management by the UN World Tourism Organisation or any other transnational tourism association.

A MULTI-SECTORAL CRISIS IN A SINGLE COMPANY

On 23 September 2019, one of the most iconic companies in global tourism collapsed after a corporate history in tourism that began in 1841 and spanned 178 years. The collapse of Thomas Cook involved a vertically integrated company, incorporating most key tourism sectors: an airline, hotels and resorts, coaches, travel agencies, ferries, tour operators, financial services and travel insurance, operating under a single corporate umbrella. The Thomas Cook collapse involved a number of dimensions including the consumer dimension, the staff dimension, the business stakeholder dimension, the financial management dimension and the destination dimension. This case is discussed in detail in Chapter 11.

The fundamental cause of the Thomas Cook collapse resulted from the accumulation of unsustainable levels of debt. By September 2019, lending institutions refused to accept more debt. Consequently, investors and shareholders were no longer prepared to provide a financial prop for the company. Associated with the level of debt was the fact that within the vertically integrated structure, some Thomas Cook divisions performed profitably and supported some divisions that lost money. The management failure centred on the failure to conduct a full and objective audit, a business restructure and address selling or divesting itself of the under-performing divisions within Thomas Cook's corporate structure.

The only dimension of risk management which benefited Thomas Cook's customers (but not its management and staff) was the consumer protection dimension, for which the company could claim no credit. In the UK, consumers who booked package tours with Thomas Cook were covered by the government backed and consumer funded Air Travel Organiser's Licence (ATOL) scheme (2020), which has operated in the UK since 1973, and the European Package Tour Directive (2020), which has operated in the European Union (EU) countries since 2015 and was upgraded in 2018. As a result, 600,000 Thomas Cook British and EU customers were repatriated or refunded for the disruption or cancellation of their holidays.

SUMMARY AND CONCLUSION

The recovery of a tourism destination or enterprise from a crisis requires a clear definition of what constitutes a crisis. The key issue in defining a crisis is the emphasis on severity. In the language of risk management, the consequences of a crisis episode represent a serious threat to the viability and reputation of the affected destination, sector or enterprise. It is also important to appreciate that while many crises in tourism result from factors external to the management of an enterprise or destination, some do arise from gross management failure. The development of social media and 24/7/365 news cycles on all media platforms has resulted in crises arising at both the finite and perceptual level.

Excellence in risk management involves the development of a proactive strategy focused on the prevention of a wide range of potential crises. Risk management policies and practices are increasingly emerging as legal requirements for many enterprises in a growing number of jurisdictions. Inculcating risk management into enterprise management will aid the ability of enterprises and destinations to effectively respond to those crisis events that could not be prevented.

The recovery of a tourism destination or enterprise involves restoring damage to technology or infrastructure, but also restoring the trust and reputation of the enterprise or destination to all stakeholders and especially consumers. The rebuilding of trust involves open and transparent communication, and when possible, first-hand demonstration of improvement measures being undertaken. This involves extensive engagement with stakeholders and with the various media platforms, which subject to the nature of their coverage to their audiences can either enhance or damage perceptions of tourism enterprises or destinations. The guiding principle of recovery is a restoration to a higher standard than the pre-crisis period – building back better. This applies equally to both the tangible and intangible elements of the tourism experience.

Approaches to risk, crisis and recovery management in tourism are subject to considerable sectoral variations. Sectors of the tourism industry, notably aviation and cruising, that adhere to global governance and regulations, are able to introduce sector-wide global risk and crisis management policies, operational approaches and protocols. However, most sectors of the tourism industry involving the majority of tourism enterprises are not subject to globally centralised sectoral governance. The growing market expectation for tourism destinations and businesses to demonstrate a commitment to safety and duty of care is likely to be the driver for a growth in the number of tourism destinations and enterprises to treat this dimension of management as vital rather than peripheral.

Discussion Questions

1. Apply the concept of destination resilience to a destination of your choice.
2. There are many possible definitions for a tourism risk and a tourism crisis. How would you define tourism risk and tourism crisis, and apply your definition to a specific destination or tourism business?
3. Perception is a major element on determining the severity of a tourism crisis. Discuss an example of a crisis in which the perception of the 'crisis' is, or was, worse than the actual situation and how would you address this in your recovery marketing?
4. This chapter featured Japan's remarkable tourism resurgence after the 2011 Tohoku earthquake and tsunami. Discuss five key elements you can identify as having facilitated the resurgence and feel free to include issues not discussed in the chapter.
5. Choose an example of any tourism destination or business of your choice and discuss at least five actions or strategies that can turn a tourism crisis into an opportunity.

REFERENCES

Air Travel Organiser's Licence (ATOL) (2020) www.caa.co.uk. Accessed 20 December 2020.

Asia-Pacific Economic Cooperation (APEC) (2017) *Strengthening Tourism Business Resilience Against the Impact of Terrorist Attack*. Report of the APEC Counter Terrorism Working Group's Workshop on 9–10 May 2017, Bali, Indonesia. APEC Counter-Terrorism Working Group, APEC Secretariat: Singapore. pp. 94.

Avraham, E. (2015) Destination image repair during crisis. Attracting tourism during the Arab spring. *Tourism Management*. Vol 47. April 2015. 224–232

Avraham, E. and Ketter, E. (2008) *Media Strategies for Marketing Places in Crisis. Improving the Image of Cities, Countries and Tourist Destinations*. Elsevier: Amsterdam. pp. 78–97.

Avraham, E. and Ketter, E. (2016) *Tourism Marketing for Developing Countries*. Palgrave Macmillan: Basingstoke, UK. pp. 16–34.

Beirman, D. (2003) *Restoring Tourism Destinations in Crisis – A Strategic Marketing Approach*. Allen & Unwin: Crows Nest (Sydney), Australia. pp. 3–12.

Beirman, D. (2014) The impact of volcanic ash clouds in 2010 and 2011 on global tourism mobility and emergency intermodal transport policy (Full Paper). *CAUTHE Conference 2014 Proceedings*. University of Queesland, Brisbane. pp. 71–82.

Beirman, D. (2016) (Updated 2019) *Tourism Risk, Crisis and Recovery Management Guide*. Council of Australian Tour Operators: Sydney, Australia.

Beirman, D. (2018) *The Impact of the October 2017 Las Vegas Massacre on Global Approaches to Hotel Security*. Working Paper CAUTHE Conference.

Beirman, D. and Van Walbeek, B. (2011) *Bounce Back – Tourism Risk, Crisis and Recovery Management Guide*. Pacific Asia Travel Association: Bangkok (also published in Japanese and Mandarin).

Beirman, D., Upadhayaya, P., Pradhananga, P. and Darcy, S. (2018) 'Nepal earthquake in the aftermath of the April/May 2015 earthquake and aftershocks. Repercussion, recovery and the rise of new tourism sectors', *Tourism Recreation Research*, 43(4): 1–11.

Benoit, W.L. (2018) 'Crisis and image repair at United Airlines. Fly the unfriendly skies', *Journal of International Crisis and Risk Communications Research*, 1(1): 11–26.

Calgaro, E., Lloyd, K. and Dominey-Howes, D. (2014) 'From vulnerability to transformation: a framework for assessing the vulnerability and resilience of tourism destinations', *Journal of Sustainable Tourism*, 22(3): 341–60.

Citonot, L. (2013) 'Thailand tourism leaders press for new safety and security measures', *Travel Daily News (Thailand)*, 11 August 2013, p. 1.

Cochrane, J. (2010) 'The sphere of tourism resilience', *Tourism Recreation Research*, 25 (2): 173–85.

Cordato, A. (2020) *Memo to the Travel Industry. Update your Terms and Conditions*. Legalwise Seminars, Sep 2030. https://legalwiseseminars.com.au/insights/memo-to-the-travel-industry-update-your-terms-conditions-after-covid-19/ Accessed 02 May 2021.

Dunn, C.E. (2012) Planning Work Health and Safety. *An Introduction to Best Practice*. CCH: Australia. pp. 205–61.

Dwyer, L., Edwards, D. and Mistilis, N. (2009) 'Destination and enterprise management for a tourism future', *Tourism Management*, 30(1): 63–74.

European Package Tour Directive (2020) https://ec.europa.en/info/law-topic/consumers/travel-and-timeshare-law/package-travel-directive_en. Accessed 20 December 2020.

Government of Nepal, (2020) *Nepal Tourism Statistics 2019*. Ministry of Culture, Tourism & Civil Aviation. p. 20

Hall, C.M., Prayag, G. and Amore, A. (2018) *Tourism and Resilience – Individual, Organisational and Destination Perspectives*. Channel View Publications: Bristol, UK. pp. 32–60.

Hampton, J.J. (2014) *Fundamentals of Enterprise Risk Management*. Amacon: New York, USA. pp. 23–40.

Hollway, S. (2007) *The 10 Commandments of Tourism Recovery*, APEC Tourism Working Group Conference, keynote address, May 2007. Surfers Paradise: Queensland, Australia.

Hong Kong Tourism Board (2021) https://discoverhongkong.com. Accessed 30 March 2021.

Insurance Council of Australia (2019) *Explaining Insurance*. Presentation by Laurie Ratz to UTS students, May 2019.

International Civil Aviation Organisation (ICAO) (2009) Contingency Procedures Against Hurricanes and Volcanic Ash is the CAR/NAM regions. ICAO Armenia, Columbia. 27 May 2009.

International Hotels and Restaurants Association (IHRA) (2020) www.ih-ra.org. Accessed 1 December 2020.

International Organization for Standardization (ISO) (2009) *ISO-31000-2009 Risk Management*. https://infostore.saiglobal.com/preview/is/2009/iso31000-2009-pdf?sku=13949. Accessed 19 December 2020.

International Organization for Standardization (ISO) (2018) *ISO-31000-2018 Risk Management*. https://theirm.org/media/6884/irm-report-iso-31000-2018-v2.pdf. Accessed 1 December 2020.

Jamal, T. and Stronza, A. (2009) 'Collaboration theory and tourism practice in protected areas', *Journal of Sustainable Tourism*, 17(2): 169–90.

Japan National Tourism Organisation (2020) *Statistics*. https://statistics.jnto.go.jp/en/. Accessed 1 December 2020.

MGM Resorts (2020) www.mgmresorts.com. Accessed 1 December 2020.

Pacific Asia Travel Association (PATA) (2015) *Report of the Nepal Rapid Response Taskforce*. PATA, Bangkok.

Pacific-Asia Travel Association (2020) www.pata.org. Accessed 30 November 2020.

Rabassa, A., Blackwill, R., Chalk, P., Cragin, K., Fair, C., Jackson, B.A., Jenkins, B.M., Jones, S., Shestak, N. and Tellis, A.J. (2009) *The Lessons of Mumbai*. Rand Corporation: Santa Monica, Cal., USA.

Ritchie, B.W. (2009) *Crisis and Disaster Management for Tourism*. Channel View Publications: Bristol, UK. pp. 4–22.

Saving of Life at Sea (SOLAS) (2020) *International Maritime Organisation*. https://imo.org/en/MediaCentre/PressBriefings/Pages/35-SOLAS-EIF-2020.aspx. Accessed 01 December 2020.

Takamatsu, M. (2014) 'The Okinawa tourism crisis management initiatives', *International Journal of Event Management Research*, 8(1): 19–33.

Tourism Authority of Thailand (TAT) (2020) *Statistics*. https://tatnews.org/tag/tourism-statistics. Accessed 1 December 2020.

UN World Tourism Organisation (2011) *Toolbox for Crisis Communications in Tourism. Checklist and Best Practice*. UNWTO: Madrid, Spain.

Wearing, S., Beirman, D. and Grabowski, S. (2020) 'Engaging volunteer tourism in post-disaster recovery in Nepal', *Annals of Tourism Research*, 80: 102802.

Wilks, J., Pendergast, D. and Leggat, P. (2006) *Tourism in Turbulent Times – Towards Safe Experiences for Visitors*. Elsevier: Amsterdam.

World Bank (2014) *Risk and Opportunity. Managing Risk for Development*. The World Bank: Washington, DC. pp. 53–74.

World Travel and Tourism Council (WTTC) (2020) *Protocols in the New Normal*. www.wttc.org. Accessed 30 November 2020.

Yates, M. (2006) 'Project Phoenix: A benchmark for reputation management in travel and tourism', in Wilks, J., Pendergast, D. and Leggat, P. (eds), *Tourism in Turbulent Times: Towards Safe Experiences for Visitors*. Elsevier: Amsterdam. pp. 263–76.

2

GOVERNMENT AND TRAVEL INDUSTRY APPROACHES TO TOURISM CRISES AND RECOVERY

INTRODUCTION

The primary focus of this chapter is to discuss the critical role of government in establishing many of the regulatory parameters for dealing with tourism risk, crisis and recovery. It will also examine key issues in risk, crisis and recovery as they apply to key sectors of the global tourism and hospitality industries. The reason for doing so is that the risk environment of aviation or cruising is quite different from that of the accommodation, conference and convention or travel retailing sectors.

The vast tourism and hospitality industry is essentially the sum of its sectoral parts. While there is a high level of collaboration between the sectors in delivering a tourism service or experience, each sector has its own specific set of risks. An understanding of the risks applying to individual sectors contributes to a holistic view of tourism risk, crisis and recovery.

One of the main challenges for tourism enterprises and destinations is to look beyond their own sectoral silos. One significant opportunity engendered by a major crisis such as the Covid-19 pandemic of 2020 is that there is no longer an option. An all-encompassing global crisis, affecting every sector of tourism, has forced even the most recalcitrant tourism industry manager to seek solutions, extending beyond their own sector. This reinforces the Mandarin language definition of crisis as 'danger and opportunity'.

GOVERNMENT AND GLOBAL/TRANSNATIONAL TOURISM ASSOCIATIONS

The outbreak of the Covid-19 pandemic during 2020 sharply focused the importance of government in defining the nature of a crisis, establishing regulatory approaches in response to a crisis and providing financial assistance to industries, enterprises and their employees affected by a crisis. As discussed in Chapter 3, which focuses on the Covid-19 crisis, governments have imposed regulations and restrictions that severely affect all sectors of global tourism.

There is a traditionally close relationship between the tourism industry and government. The majority of tourism marketing boards at either national, state/provincial/prefectural or local level are funded by one of the three tiers of government: national–regional–local. Even at the transnational and global level, governments have a significant involvement in tourism. The UN World Tourism Organisation (UNWTO), International Maritime Organisation (IMO) and International Civil Aviation Organisation ICAO) are all UN bodies. Transnational tourism bodies, include the European Travel Commission, Caribbean Tourism, ASEAN Tourism and the APEC Tourism Working Group, funded by government instrumentalities with private sector support (Cooper, 2016). It was widely recognised, until 2020, that tourism served the economic goals of most governments. International air and sea routes, between countries, are determined via a range of bilateral agreements between relevant governments. Governments determine the rules and conditions of international tourism and visitor entry into each country through visas.

The World Travel and Tourism Council (WTTC) is a global association representing the largest 150 private sector companies in global tourism across all sectors. Its members include leading hotel chains, airlines, cruise companies, attraction companies, convention and meeting operators, tour operators, travel agencies and online booking companies. The WTTC commissions, ongoing research on global tourism trends and the wealth of its members provide it with significant influence in lobbying government to support the interests of the tourism industry and especially its members.

During the April 2017 WTTC's Global Tourism Summit, held in Bangkok, all attendees recognised that risk, crisis and recovery management, along with sustainability and an embrace of modern technology represented the three primary issues for the global tourism industry. An entire session of the two-day summit was devoted to discussing approaches to tourism safety, crisis and business continuity and recovery management. This was not the first time that global tourism associations (including the WTTC) had acknowledged the significance of risk, crisis and recovery. In 2012, the WTTC's Global Tourism Summit held in both Sendai and Tokyo focused heavily on Japan's recovery from the 2011 Tohoku earthquake and tsunami, which struck the Sendai region of Japan.

It was appropriate that the keynote speaker for the 2017 crisis management session, Robert Jensen, is CEO of the world's oldest privately run travel crisis management company, Kenyon International, established in 1906. Kenyon provides a comprehensive range of crisis management services for many of the world's airlines. Many destination management organisations contract Kenyon to manage both tourism crisis and recovery management programmes. In 2002, the International Air Transport Association closed its crisis management unit and outsourced it to Kenyon International Emeregency Services. The activities of Kenyon include the provision of risk management and crisis communications training for its clients, tailored recovery strategies, assistance and counselling to victims and their families, mortuary, funeral and cremation services, legal assistance, identification of victim possessions, medical treatment and business/destination reputation and recovery management among many other services. It is noteworthy that Kenyon began its life as a business prior to 1906 as an undertaker company and its evolution to tourism crisis management was the result of a major US train crash in 1906, which led to it taking on a far broader dimension of crisis management than the traditional undertaker's role (Kenyon International Emergency Services, 2020).

The private sector plays a significant role in tourism risk, crisis and recovery. The world's largest tourism oriented risk management company, International SOS (2020) has offices in over 50 countries and employs over 14,000 people. Many thousands of travel agents worldwide and their corporate tourism clients outsource International SOS and other private companies offering similar services, to manage their risk management, tourism security and duty of care policies. The higher education sector, along with government and privately funded colleges and research centres have played an important role in providing much of the research data and strategic directions, which identify and describe best practice approaches to support tourism industry and destination resilience. Many larger travel companies and destination management organisations from all sectors of the tourism industry have established risk and crisis management units, which devote their full attention to the development and implementation of risk and crisis management policies, protocols and practice.

In 2011, the author, in association with Thailand based tourism crisis management consultant Dr Bert Van Walbeek, principal of a private tourism crisis management consultancy the Winning Edge, co-authored the Pacific Asia Travel Association's (PATA) tourism risk, crisis and recovery manual (*Bounce Back*), which was used by over 50 Asia Pacific tourism ministries as their risk, crisis and recovery management template (Beirman and Van Walbeek, 2011). The manual was reinforced by a series of webinars presented under the auspices of PATA in 2012–2013. This manual built on an earlier manual produced by Dr Van Walbeek for PATA in 2003, *Crisis, It Won't Happen to Us* (The Winning Edge, 2003). In 2020, the many new challenges to tourism, which arose from the Covid-19 pandemic, led PATA to rewrite its crisis management and crisis communications guides (PATA, 2020).

The 2017 WTTC summit held in Thailand's capital, Bangkok, attracted over 1,000 delegates (including the author) and involved the active participation of virtually all leading global and transnational tourism associations, including the UNWTO, PATA, the International Air Transport Association, Cruise Lines International and the world's major hotel chains. Full WTTC membership is by invitation only and members are sought from the 150 largest companies in global tourism from all key sectors of the industry. The industry-wide acceptance of the significance of tourism risk, crisis and recovery, reinforced at the 2017 summit, is a relatively recent phenomenon. In fact, the 21st century has marked the awakening of the global tourism industry to its importance. The April 2012 WTTC Global Summit, held in Tokyo and Sendai marked symbolic support of the WTTC for Japan's tourism recovery, a year after the 2011 Tohoku earthquake and tsunami.

As discussed in Chapter 1, the diversity of the global tourism industry into multiple sectors of activity resulted in some sectors of the tourism industry giving far more primacy to risk, crisis and recovery management than others. The governance of risk, crisis and recovery management has also varied significantly in the sectors.

SELECTED SECTORAL APPROACHES AND RESPONSES TO RISK AND CRISIS

Cruising

Some key sectors of the global tourism industry have recognised the significance of risk, crisis and recovery management for many years. In 1914 the IMO, which covers all merchant and passenger shipping lines worldwide, approved the first draft of the IMO treaty Safety of Life at Sea (SOLAS, 2020), which remains the IMO's guiding principle concerning safety and risk management. SOLAS was a direct shipping industry response to one of history's most infamous maritime disasters, the 1912 Titanic sinking. A significant contributing factor to the high loss of life arising from the Titanic disaster was the shortage of lifeboats – one of the first issues addressed by SOLAS.

Although SOLAS has undergone multiple amendments since 1914, the SOLAS protocols expressed a broad commitment by the cruise sector to address the safety and security of passengers and crew at sea. The IMO is a UN body responsible for the global governance of safety and operational standards for global shipping, including the cruise industry. The IMO is committed to high standards of safety, maintenance, health and hygiene, for all global shipping. The IMO's commitment to shipping is reinforced by Cruise Lines International Association, (CLIA), the global association of the cruise industry.

Cruising, as a sector within transport, is readily amenable to global governance especially in the fields of risk, crisis and recovery management and it represents a core priority shared by all cruise companies. In 2019, over 30 million people engaged in cruising (CLIA, 2020). This number does not include the far larger number of passengers who use ferries and other short haul seaborne transport. For the sake of clarity, cruising involves sea ships that provide transport, passenger accommodation, a full food and beverage service and entertainment while travelling on a route encompassing either sea travel or a range of destinations in a set itinerary. By contrast, some ferries (especially those that travel between countries on long haul services) may provide some of the above services but their primary role is the provision of point-to-point maritime transport. Although commercial cruising operates in many forms, including ocean and sea cruising, river cruising, adventure cruising and small ship cruising, to name a few, the majority of cruise brands are dominated by a relatively small number of companies. The multi-billion dollar costs involved in cruising operations and the construction and maintenance of massive and complex vessels requires those companies engaged in the sector to be highly capitalised.

The world's two largest cruise companies, Carnival International and Royal Caribbean, own 20 of the world's leading brands between them. These include widely recognised brands such as P&O, Cunard, Princess Cruises and Celebrity Cruises. In 2019, the world's largest cruise ships approached a size where a single vessel was on the verge of being able to accommodate over 10,000 passengers and crew. From a risk management perspective, operating a vessel of this size with such a large number of passengers is accompanied by a significant range of risks, including food and alcohol services, medical care, crime prevention, mechanical failure of the vessel and its systems and weather-related risks to name a few. In the language of risk management, the likelihood of a massive cruise ship sinking in mid-ocean is small but the consequences would be potentially catastrophic.

Until recently, the major cruising companies were American or European corporations servicing a predominantly Western market. Until 2016, almost 50 per cent of global cruise passengers were US citizens and most of the remainder were British, European and Canadian. In recent years, cruising has grown in popularity in other parts of the world, notably Australia, which experienced better than ten per cent annualised growth, between 2004 and 2018. In 2018, 1.4 million Australians cruised, representing five per cent of the 28 million passenger global cruise market during that year. CLIA's former CEO Cindy D'Aoust forecasted that the bulk of global cruising's future growth would come from China, Japan, South Korea and South-East Asia. The rise of Asian outbound tourism was a global phenomenon, which applied to all sectors of tourism. The large number of orders for new cruise ships and forward bookings for cruises originating from East and South-East Asian ports until the end of 2019 suggested that that the CLIA's forecasts were correct until the onset of Covid-19 in 2020, which proved especially damaging to the reputation of the cruise sector.

The ultimate risk for a cruise operator is the sinking of a fully laden vessel on the high seas by either natural causes, or through human intervention or error. Lesser but still significant threats apply to contagious diseases, which at worst can lead to the rare phenomenon of a 'plague ship', mass outbreaks of contagious illnesses and food poisoning. Cruise ships have been a potential target of terrorism since the hijacking of the Italian cruise ship, Achille Lauro by terrorists from the Palestine Liberation Front in 1985.

Maritime choke points are narrow straits linking oceans and seas on major sea trade routes (see Figure 2.1). They include the Gulf of Aden, (also known as the Bab el- Mandeb straits) at the southern entrance of the Red Sea, the Straits of Hormuz (linking the Persian Gulf and the Indian Ocean), and the Straits of Malacca where Singapore, Malaysia and Indonesia meet, where there is a very significant threat of piracy, largely targeting merchant ships. In addition to piracy targeting merchant shipping, pirates have attempted to target civilian cruise ships. In recent years, the failed state of Somalia has been a major centre of piracy in the Gulf of Aden. Since 2008, navies from 20 countries regularly patrol the waters of the Gulf of Aden under the authority of the Maritime Security Patrol Area to prevent pirate attacks on shipping.

A number of heavily publicised criminal incidents including onboard murders have occurred on cruise ships. Assault, robbery and rape are among criminal risks that cruise companies have dealt with. In any community of several thousand people, especially a large, temporary, affluent community where passengers are in vacation mode, alcohol flows freely and inhibitions are shed, the motivation, opportunity and likelihood of opportunistic crime committed by passengers or crew are high.

Two of the more infamous and recent crisis incidents affecting cruising occurred in early 2012 to two vessels flying under the Costa Cruising Brand (a division of Carnival Cruising International). In both cases, the causes of the problems involved were a result of management and maintenance failure. The grounding and partial sinking of the Costa Concordia in January 2012, carrying over 4,000 passengers and crew, which ran aground on the rocks of an Italian island (Isola del Giglio) and sank in shallow water, was caused by poor captaincy and navigation. The relatively small number of casualties, 33 dead (32 passengers and one salvage worker), was largely due to the fact that the ship was literally grounded and the majority of passengers could easily evacuate the ship.

The grounding of the Costa Concordia was followed in February 2012 by the complete mechanical failure of another Costa liner, the Costa Allegra in the Indian Ocean near the Seychelles. Although there were no deaths or serious injury resulting from the mechanical failure, the ship required towing to the nearest point. These two incidents within the space of just over a month inflicted severe damage to Costa Cruising's brand and its reputation. The reputational damage to Costa was exacerbated by the poor response of Costa's senior executives to both incidents. These incidents are fully discussed in Chapter 11.

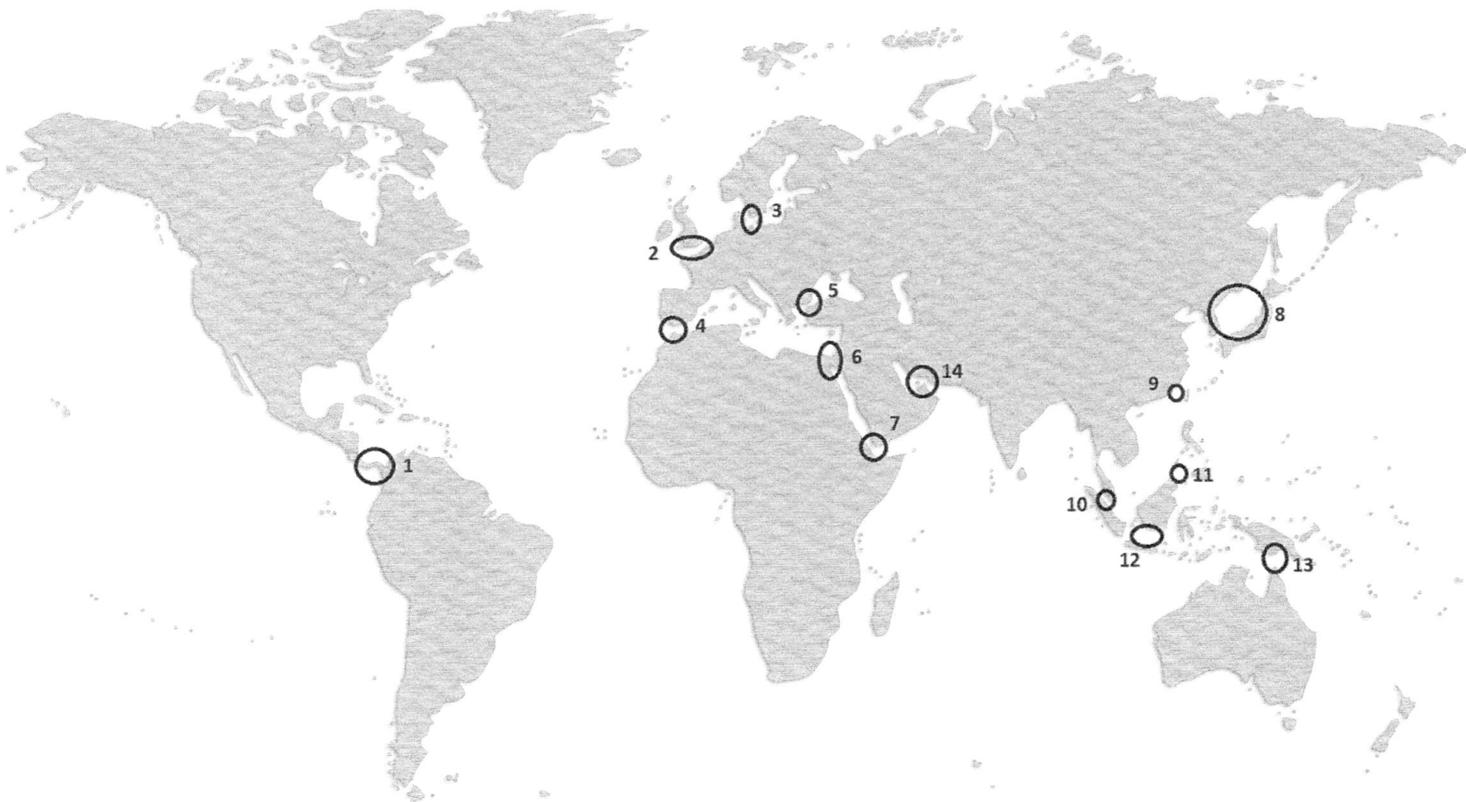

Figure 2.1 Maritime choke points

Source: Image by Pete Linforth from Pixabay.

Key: 1. Panama Canal, 2. English Channel, 3. Entrance to the Baltic Sea, 4. Straits of Gibraltar, 5. Bosphorus Strait, 6. Suez Canal, 7. Bab el-Mandeb, 8. Sea of Japan, 9. Strait of Formosa/Taiwan Strait, 10. Strait of Malacca, 11. Sulu Sea, 12. Java Sea (an active site for global piracy), 13. Torres Strait, 14. Strait of Hormuz

In the event of management or equipment failure, the golden rule in a post-crisis recovery is demonstrable concern for the welfare of victims, accompanied by a clearly communicated statement from management that failures are being fully addressed by the company.

Since the Costa Concordia and Costa Allegra incidents of 2012, the parent company Carnival Cruising and other cruise companies gave far greater emphasis to crisis communications. CLIA as the global association representing the cruise industry has introduced training programmes on risk and reputation management, with a focus on the training of senior management of cruise lines in effective recovery strategies in response to negative incidents that occur on cruises from time to time.

The crises experienced by the cruise sector resulting from Covid-19 outbreaks on cruise ships are discussed in Chapter 3. However, the cruise sector suffered significant reputational damage related to Covid-19 outbreaks on cruise ships. While the leaders of CLIA bunkered down and generally avoided engagement with the media, the mass media had a field day savaging the reputation of the cruise industry. Ironically, the CLIA and cruise company executives actively engaged with travel agents and the travel industry media. Although the CLIA has sophisticated PR resources and well-qualified spokespeople to respond to many of the negative allegations, and in some cases, inaccurate reports about cruise companies that appeared in various news media outlets, they frequently have a baffling reluctance to use them, especially when they are invited to appear on TV current events programmes. In September 2020 the CLIA finally released an agreed set of Covid-19 protocols, which they released to all stakeholders including the media.

Aviation

The airline sector, in common with cruising, has demonstrated a longstanding concern with safety and risk management. The 1944 Chicago Convention, ratified by the establishment of the International Civil Aviation Organisation in 1947, essentially commits the global civil aviation sector to a wide range of specified standards for safety, airline operations and maintenance (ICAO, 2020). These standards apply to all airlines operating under the authority of all UN member states. The ICAO was one of the earliest UN agencies. The International Air Transport Association (IATA) is an industry association, supported by 85 per cent of the world's airlines who are members. IATA sets even higher safety standards for its members than the ICAO.

Until 2002, IATA had its own crisis communications unit, which focused on developing and implementing a series of protocols for risk management and the managing and recovering from crisis events. This unit was on call (for a fee) to any IATA member airline that lacked its own crisis management unit. After 2002, IATA outsourced its role in crisis recovery to Kenyon International. Many of the world's largest airlines, including most full-service (legacy) carriers and the larger low-cost carriers (such as Air Asia), maintain their own crisis management and crisis communications units. The growth of multi-branded airline consortia in recent years, including Oneworld, Star Alliance and SkyTeam, enabled those consortia to establish risk, crisis and recovery units servicing their member airlines.

The 9/11 attack on 11 September 2001 represented the most significant crisis for the aviation and airport sectors in recent history, until the Covid-19 outbreak of 2020. Civil aviation hijackings have been a recognised threat to aviation since the 1960s. However, the unique dimension of the 9/11 attack involved tasking a commercial airline as a weapon of mass destruction. This was done with devastating effect through hijackers taking control of fully fuelled commercial aircrafts, flying into and destroying the World Trade Center in New York City and inflicting severe damage to the Pentagon in Washington, DC (Beirman, 2003).

Although the targets of the 9/11 attacks were specific and localised, the impact on the aviation sector was global. Between September 2001 and June 2002 there was a global decline in aviation bookings and passenger numbers. The 9/11 attack forced a global, full-sectoral re-evaluation of security protocols and measures applying to both airlines and airports. The full review of safety and security protocols, completed by late 2002, involved the joint participation of IATA, ICAO and Airports Council International (ACI) (the global association for airports). One of the simplest and most obvious measures agreed was ensuring that cockpit doors in all civil aircraft were locked and hardened. Although many of the revised global protocols could not be made public, the fact that a new and greatly enhanced set of security protocols were implemented provided a strong basis for the airline and airport sectors to engage in a concerted recovery campaign during 2002 and 2003. The core intention of the campaign was to restore consumer confidence and a broad range of business and government stakeholders in the safety of aviation.

In recent years, the aviation sector acted collectively in response to a range of shared challenges. In April 2010, the eruption of the Eyjafjallajokull volcano in Iceland led to the closure of most airports in Western Europe and the UK and the cancellation of thousands of flights due to concerns about volcanic ash clouds. Over a five-day period, the ash cloud event disrupted travel for an estimated ten million people. The flight cancellations caused massive disruption to flights within, to and from Western Europe. The ICAO's justifiably cautious approach to flight operations involving volcanic ash clouds conformed with its ash cloud protocols, published in 2009, developed in response to earlier volcanic ash cloud incidents (Beirman, 2014).

These incidents demonstrated that volcanic ash particles above a given density had the potential to immobilise jet engines with catastrophic results. Adherence to ICAO protocols (ICAO, 2009) called for a grounding of airlines operating in areas of high-density volcanic ash at cruising altitudes. This meant that casualties, which may have resulted from aircraft crashing when flying through clouds of volcanic ash, were averted. Many individual carriers determine their own set of volcanic ash cloud protocols, stricter than those of the ICAO. It is notable that there was a high degree of cooperation between government transport ministries and other relevant government authorities, airport managers, air traffic controllers and the airlines in determining the advisability and safety of airline operations during volcanic ash cloud events.

During the April 2010 European ash cloud event, several thousand long-haul passengers were stranded at stopover points, en-route to Europe from points of origin such as East Asia and Australasia. They received an unscheduled lesson about the difference between full-service IATA member carriers' and low-cost non-IATA member carriers' approach to duty of care. IATA has long-standing rules, applicable to passengers who experience stopover delays of over eight hours beyond the control of the carrier. In such cases, the airline is obliged to accommodate and feed affected passengers. Although such passengers experienced disruption and delay they tended to do so in relative comfort. Airline passengers stranded in Bangkok at that time also had to contend with a violent political upheaval, which gripped the city. Passengers travelling on low-cost, non-IATA airlines, which have no obligations to service or compensate delayed passengers, were literally left to their own devices at their own expense, sitting out delays of up to a week at airports in Asia and the Middle East.

In terms of business recovery, airlines prefer to apologise, and in some cases, financially compensate their customers for a delayed or disrupted trip than to risk loss of life. Subsequent volcanic ash cloud events in Chile in June 2011 (which affected much of South America, New Zealand and Southern Australia) and Bali in 2017 and 2018 have seen airlines respond to volcanic ash clouds in accordance with ICAO protocols, which require grounding of commercial airline flights in areas affected by volcanic ash clouds.

Airlines are well aware of the linkage between risk/crisis and reputation. Although the world's airlines are statistically the safest mode of transport, the consequences of an airline disaster can be both severe and enduring. In 2014, Malaysian Airlines suffered two disasters damaging the carrier's reputation long after the event. The disappearance of MH370 carrying 270 passengers and crew, during a flight between Kuala Lumpur and Beijing in March 2014 remains one of aviation's most baffling unsolved mysteries. The shooting down of MH17 by an anti-aircraft missile fired from Ukraine, while flying between Amsterdam and Kuala Lumpur in July 2014, resulted in the death of 283 passengers and crew. Malaysian Airlines was subject to media criticism for flying over an area subject to military conflict. However, other carriers including Thai Airways and British Airways were flying over the same flight path at the time of the incident.

While the fundamental cause of the MH17 disaster is not in dispute, neither Russia nor Ukraine (nor the various guerrilla militias) have accepted responsibility for firing the fatal missile. Malaysian Airlines' reputational recovery from the impact of these two has been made exceedingly difficult by the fact that the cause of both incidents is not fully resolved and closure has not been achieved. In 2019, several people were charged with culpability for the missile which downed MH17 over Ukraine although no government or militia group has accepted responsibility. The disappearance of MH 370 remains a mystery. In both cases, the company was widely deemed to have been at fault and consequently Malaysian Airlines undertook significant management changes from 2014 as part of its post-crash recovery programme.

Commercial aircraft are complex pieces of machinery and the failure of a single part among millions has the potential to be fatal. One of the great risks for airlines and aircraft manufacturers involves the failure of vital components. Boeing, one of world's two leading commercial aircraft manufacturers, incurred significant commercial and reputational damage following two crashes of the 737 MAX, one operated by Lion Air (Indonesia) in October 2018, which killed 189 passengers and crew, and the other operated by Ethiopian Airlines in March 2019, which killed 157 passengers and crew. The subsequent grounding of all 737 Max aircraft pending extensive checks for re-licensing has had a negative impact on Boeing's reputation. The company was forced to re-negotiate many orders with the world's major carriers. Boeing has fully cooperated with the regulatory authorities to ensure the defects to the 737 Max are fully rectified.

Airlines and cruise ships are both providers of mass transport and whilst both modes have a common commitment and a generally high level of safety, an aircraft crash or the sinking of a ship has the potential to be mass casualty disaster. In both modes there is a high degree of global risk and crisis management governance and increasingly common approaches to recovery in the event of a crisis event. The aviation and cruise sectors at both the sectoral and individual business level train their management and staff in crisis communications and there is a high degree of understanding that restoring confidence after a crisis event is central to both sectoral and corporate reputation. It is important to understand that integral to effective risk management of these sectors there is a high level of cooperation between the sectors and government-run emergency services, including police, ambulance, rescue services and in some cases the military.

The outbreak of the Covid-19 pandemic in 2020 has led to a new crisis challenging the viability of global airlines. The combination of worldwide international and in some cases intra-national border closures, and to a lesser extent, social distancing regulations, have literally grounded most of the world's commercial passenger airlines. At best, international and domestic airline routes have been severely restricted. According to IATA, flight movements in April 2020 were 97 per cent down on the number of flight movements in April 2019 (IATA, 2020). As international flights are subject to bilateral agreements negotiated by governments, airlines are captive to governments agreeing to re-open flight routes between countries. The 2020 pandemic has emerged as the most severe crisis experienced by civil aviation since World War II. More detail on this issue is in Chapter 3.

Rail and Ground Transport

All modes of transport have high levels of risk awareness, crisis and recovery management protocols. In the case of rail, risk and crisis management governance occurs at the national level, apart from Western Europe where the European Union seeks to exercise governance at the transnational level. The rise of high-speed rail networks, especially in Western Europe, China, Japan, Korea and Taiwan has made this mode of transport highly competitive with short-haul airline services but the quality of the rail infrastructure to support it varies from country to country, and there have been some cases of disastrous train crashes.

The level of risk applicable to coach and automobile transport is far greater than for airlines and cruises. Road transport, whether it involves tour coaches, off road vehicles, rental cars, taxis or the more recent growth of Uber, is subject to governance which varies within an individual country. Road transport operators generally operate on public roads, which involves exposure to other drivers of variable skill. States and provinces within a single country frequently impose variable conditions and standards for vehicle registration and licensing of drivers. This results in some jurisdictions imposing stricter or laxer regulations than others. The vast number of private and commercial road users and the variability of road conditions magnify the level of risk applying to road transport compared to other modes of transport, which operate on relatively controlled air, sea and rail paths.

Most companies engaged in tourism-related road transport generally seek to maximise the safety and maintenance of their vehicles. However, in some jurisdictions where governance is lax, some companies cut costs by minimising vehicle maintenance to maintain the price competitiveness of their services. In turn, these cost-cutting practices exacerbate risk.

The recent rise of the tourism transport disruptor, Uber, has upset many traditional providers of taxi services that are generally subject to expensive government-imposed drivers' and operators' license and vehicle registration fees. Uber has employed a common legal loophole in which Uber vehicles are defined as private vehicles, engaged in commercial business as opposed to tourism vehicles providing public transport. Consequently, Uber has largely avoided the expensive licensing overheads and rigid safety requirements applying to taxis and hire cars. From a risk management perspective, this increases risk for passengers and drivers. However, consumer tolerance for this increased level of risk leads to reduced costs for drivers and their passengers.

Motor vehicles and bicycles remain significant providers of tourism-related transport. Vehicles used for tourism range from self-drive rental vehicles to passenger transport such as coaches, taxis or customised off-road vehicles. While the likelihood of a major vehicle accident or mishap is considerably greater than it is for any other mode of transport, the consequence of a vehicular mishap is far less than would apply for a ship sinking, airline crash or train wreck. Nevertheless, a serious mishap involving a fully laden tour coach could potentially cost more than 50 lives and any mishap involving a vehicle used for tourism has the potential to damage the reputation of the operator or company.

The shared challenge for all companies in the transport sector is to proactively minimise and effectively treat risk. All transport-related companies should be aware that disasters can and do occur. Whether they are a small taxi firm, an international airline or cruise operator, personnel well trained in crisis communications and crisis management should be mobilised to lead the management of a crisis event and be the point of communication with all relevant stakeholders, including the media. Many companies, including those with crisis management teams, experienced reputational damage when they failed to honestly and sensitively communicate with relevant stakeholders. Victims and their next of kin should be the top priority, especially in those incidents that involve casualties. Crisis communications plays a major role in the recovery and reputation management process. The response of companies in the first hour following an aberrant incident is critical. Businesses must now assume that through social media operating on hand-held devices a negative event can become a global news event within seconds.

The key element of effective response to a crisis event is timing. If a contingency plan has been developed for a business or destination, the company should have developed a strategic plan enabling it to respond almost immediately. The first hour after an event has occurred, be it an act of man or an act of nature, is the most critical. There is a need to establish to stakeholders and the media that the business operation is aware of events and is in the process of managing the situation, even if all pertinent facts are not immediately to hand.

Figures 2.2 and 2.3 (Beirman and Van Walbeek, 2011) summarise the core elements of an effective response to a crisis event irrespective of whether the crisis results from management failure or external circumstances beyond management control. One area in which the transport sector has led other sectors of the tourism industry has been in the development of cooperative relations with emergency service providers.

So far, in this chapter, the core emphasis of sectoral approaches to tourism recovery and resilience has focused on the transport sector. All sectors of the tourism industry share a common concern about the post-crisis recovery and enterprise resilience. However, unlike the airline and cruise sectors, the attractions, accommodation, events and tour operator sectors don't operate in accordance with global risk management governance that applies to airlines, airports and cruising. At the operational and governance dimension, accommodation, providers, attractions and events are bound by risk management laws and practices which apply to the legal jurisdictions in which they operate. At an enterprise level, individual companies or chains often set standards that apply at the corporate level.

3.1 Emergency response procedures	3.1.1	Leadership
	3.1.2	Crisis response speed
	3.1.3	Internal–External communication
	3.1.4	External suppliers handling
	3.1.5	Hospital(s)
	3.1.6	Morgues(s)
3.2 Investigation	3.2.1	Crime
	3.2.2	Legal and political
	3.3.3	Insurance
3.3 Family assistance	3.3.1	Information handling
	3.3.2	Transportation
	3.3.3	Accommodation
	3.3.4	On-site support
3.4 Communication	3.4.1	Media: Traditional print and electronic, online and online social media
	3.4.2	Hospital(s)
	3.4.3	Morgue(s)
	3.4.4	Investigation
	3.4.5	Government

Figure 2.2 Crisis response

Source: Beirman and Van Walbeek, 2011

4.1	Business continuity plan	4.1.1	Rehabilitation
		4.1.2	Normalisation
		4.1.3	Expansion
4.2	Human resources	4.2.1	Impact of reduced labour needs
		4.2.2	Motivation and training
		4.2.3	Increase crisis awareness
4.3	Debriefing	4.3.1	Thank all involved
		4.3.2	Follow up with victims
		4.3.3	Update Crisis Management Plan

Figure 2.3 Crisis recovery

Source: Beirman and Van Walbeek, 2011

The Accommodation Sector

The accommodation sector is arguably the most complex sector in the tourism and hospitality industry, especially in relation to risk management. Accommodation providers range from hotels and resorts, accommodating thousands of guests, through to private homes and apartments providing accommodation for individual guests, couples or small groups. In recent years, the rise of Air BnB, an enterprise enabling travellers to organise tourist accommodation online with private property owners, has resulted in its rapid growth. Air BnB has become the world's largest provider of tourist accommodation in terms of the number of properties and financial turnover. The accommodation sector operates under a very broad range of accommodation types, including hotel and motel rooms, serviced apartments, caravan, cabins and campsites, backpacker dormitories, guest houses, inns and pubs, luxury resorts, time share apartments, private B&Bs, religious hospices, private short-term rental apartments and homes, authentic accommodation unique to a national culture and many more.

Although there is a global association of accommodation/hospitality providers, the International Hotel and Restaurant Association (IHRA) lacks any power to either set or enforce any global standards of service, maintenance, building standards, security or risk management policies that apply to the hospitality sector. The IHRA is primarily a networking and advocacy body, lacking the governance role of the IMO or the ICAO, arguably, any attempt to impose common risk management standards on the entire accommodation sector would be confused. The sector is far too diverse to be subject to a common sectoral approach to security, risk, crisis and recovery management.

Since 2001, the accommodation sector and especially high-profile hotels in major cities and international resorts have become a target of choice for terrorist attacks. After 9/11 the hardening of security in the aviation sector (airlines and airports) led terrorists to focus on softer targets. As an integral part of the hospitality industry the traditional role of hotels are places of welcome to visitors. Major central business district (CBD) hotels welcome a range of visitors, including in-house guests, event and function attendees, shoppers, people coming to dine at the hotel restaurants and cafes or people choosing a place to meet or be noticed. In the spirit of hospitality, hotel management, staff, guests and visitors are resistant to imposing security or screening measures, which may be perceived as a breach of privacy.

However, this very concept of welcoming hospitality makes hotels compelling targets for terrorist organisations seeking to target venues where large numbers of people from multiple nationalities gather. The roll call of hotels targeted by terrorist attacks in recent years is long and bloody. Major terrorist attacks targeting tourists in hotels and resorts since 2001 have occurred in Indonesia, India, Egypt, Tunisia, Pakistan, Jordan, Mali, Burkina Faso, Kenya, Thailand, Philippines, Libya, Belgium, USA, Sri Lanka and many other countries. Terrorist attacks represent the tip of the iceberg of risks to hotel guests, which include crime, accidents and illness (especially food related).

From a sectoral perspective, accommodation providers lack the wide-ranging risk management governance, which apply globally to airlines, airports or cruising. Many accommodation providers are subject to national safety and security regulations, applicable to specific countries. A number of leading hotel chains, including the Intercontinental Hotel Group, Hilton Hotels, the Marriott Group and Accor, require hotels branded within those chains to adhere to security and risk management standards set by the chain's global management.

Attractions

The attractions sector, in common with the accommodation sector, is exceedingly variable. Many attractions including national parks, heritage sites, major galleries and museums, zoos and key historic memorials are government owned. Conversely, many artificial attractions such as amusement and theme parks are privately owned and operated. Two of the world's biggest private companies, both of which are multi-branded, are the US-based Disney Corporation and the UK-based Merlin Entertainments. Between them, these companies own many hundreds of attractions all over the world. Madame Tussauds wax model museum (part of the Merlin Entertainments group) was long regarded as a quintessentially London attraction from its opening in 1835 until 1970. In 1970, Madam Tussauds began the process of globalisation when the first non-UK Madam Tussauds was opened in Amsterdam. They now exist in 21 countries.

Attractions come in multiple forms. From the earliest days of travel and tourism, sacred sites and sites of religious and spiritual significance to the world's many religions and spiritual movements have attracted visitation and veneration. Many of these sites are owned and managed by relevant religious or spiritual institutions. A risk associated with some religious and spiritual sites are that their ownership is contested. In June 2020, the Turkish government's decision to declare the Hagia Sophia site in Istanbul as a mosque outraged Christians all over the world. Orthodox Christians have regarded the former Byzantine cathedral as one of their primary holy sites. The primacy of some holy sites in Jerusalem is a matter of dispute between Jews, Christians and Muslims for centuries. Ownership and control of the Church of the Holy Sepulchre in Jerusalem's old city (believed by Catholics and Orthodox Christians as the site of Christ's crucifixion) is claimed by seven different Christian denominations whose mistrust of each other is so great that they entrusted the keys to a Muslim family for 15 generations. Sites of scenic beauty, architectural grandeur, cultural significance and historical or archaeological significance are often commercialised and commoditised. This often contributes to the deterioration of such sites.

From a risk management perspective, attractions involve an extensive variety of risk issues. These include crowd control and control of visitor flow, carrying capacity, safe access, duty of care for visitors and staff, and the physical integrity and maintenance of the attractions. Security issues to protect visitors, the site and treasured objects within the site represent some of the risk management issues. Amusement parks have had to cope with the risks of accidents involving people killed and injured during activities and rides. National Parks and ski resorts have had to cope with accidents, forest fires and degradation arising from over-use or misuse by visitors. Some attractions have had to develop strategies in response to politically or religiously motivated violence. In common with the accommodation sector, there is no global sectoral governance of risk management in the attractions sector but attractions are required to adhere to risk management laws and protocols that apply to the governmental jurisdiction in which they apply.

Meetings, Events, Exhibitions and Conventions

There has been a symbiotic relationship between tourism, meetings, conferences/conventions and events since the earliest days of tourism. People travelling to events have relied on the transport and accommodation infrastructure to enable access, nearby accommodation and access to attend events. Events, meetings and conventions frequently leverage on the attractions of the host destination to encourage people to attend, and include pre- and post-tourism activities as a value-added element of the event itself.

In common with accommodation and attractions, meetings, events, exhibitions and conventions come in many forms and sizes. Mega events (which attract millions of people) range from global sporting events such as the Olympic Games and the World Cup football (soccer) finals to pilgrimage events such the annual Haj (Islamic pilgrimage) to Mecca and Medina in Saudi Arabia, and staged events such as the World Trade Expo, Eurovision and massive concerts. Conventions and meetings can be political, commercial, cultural, religious, military, or a spectacle to celebrate as the marking of a significant anniversary or rites of passage, such as weddings and funerals.

One of the features of mega events in more recent years has been the massive growth in expenditure in all forms of risk management, especially security. The hosting of mega events is keenly sought by countries to showcase the host city and country as a tourism destination. This hosting of events involves reputational risk applying to the event brand and the venue city/state or country. Risk management for events involves far more than a concern for security. The staging of events frequently involves the event organiser working with multiple small contractors who may have responsibility for staging, lighting, sound equipment, temporary exhibition structures and exhibitor booths and displays. Event organisers and venues have concerns about occupational health and safety, crowd management, medical care, fire prevention and duty of care.

Regrettably, the 1972 Munich Olympics is more often recalled for the terrorist act that killed 11 Israeli athletes and disrupted the games than the outstanding performances of elite athletes or the urban delights of Munich. Consequentially, officials at all succeeding Olympic Games have invested heavily in security. The core concepts of safety, security and duty of care govern extensive risk management planning in the organisation and management of meetings, events, exhibitions and conventions.

Tour Operators

Tour operators are in essence one of the most collaborative tourism industry sectors. Tour operators are involved in outbound tours (international from tourism generating region to international destination), inbound (from international source market to a tourism destination region) and domestic tour programmes. Essentially, tour operators package a tour programme that typically involves many of the principal sectors of tourism (air, land and sea transport, cruises accommodation/hospitality, events, attractions and expertise in the form of tour guides).

The risk management challenges for tour operators involve effective coordination of risk management to cover staff and customers in the smooth running of a tour with multiple elements. Tour operators also have to audit the safety and security situation that applies to each destination in which they operate, and to ensure there are backup and evacuation plans in the event of a crisis arising in those destination. They also need to ensure that contracted personnel and equipment are qualified and fully maintained. Although all but the largest tour operators contract local companies to manage much of the content of tours, in the eye of the consumer, they are deemed responsible for the overall quality of each element of the tour programme.

Travel Agents and Online Booking Agencies

Travel agents and online booking agencies coordinate travel arrangements for their clients, ranging from simple point-to-point flight bookings to complex itineraries involving tours, accommodation, flights, rail trips, event bookings, car rental and tours. Many principal service providers (notably airlines) have progressively reduced commission payments to travel agents. Agencies, which have maintained their financial viability, tend to prefer complex itineraries (especially for leisure travellers) to bookings consisting of just flights. Complex itineraries provide extra justification for the service fees travel agencies need to charge to compensate them for the lower commissions they receive from principals in recent years.

Corporate travel agents or their online equivalents, usually known as 'destination management companies', tend to focus on providing a full suite of travel services for corporate clients and their staff. In most cases, these are aligned with the duty of care policies now required by government for businesses in relation to staff or executive travel undertaken as an element of their corporate business. These include an assurance that travellers are properly insured for the duration of their business trip and that they have access to medical and other services, which may be required in the course of their travel. The choice of destination offered to corporate customers usually accords with corporate adherence to government travel advisories determined by governments in the tourism generating country.

In common with tour operators, travel agents are the point of contact when clients experience problems during the course of their booked travel. However, they may not necessarily be directly responsible for the resolution of those problems.

Some travel agents, especially those which focus on the corporate market, have taken risk and crisis management seriously and for risk averse businesses this has become a selling point. However, many travel agents have tended to adopt the approach that if one destination was problematic, there were ample alternative marketable destinations. The 2020 Covid-19 pandemic rendered this approach to risk largely untenable in many countries especially when consumer demand for international travel dropped sharply from March 2020, largely due to government-imposed travel bans and restrictions.

DESTINATION APPROACHES TO RECOVERY AND RESILIENCE

In 2011 the UNWTO's Risk and Crisis Director Dr Dirk Glaesser initiated the concept of the integration of emergency management and tourism across all sectors of the tourism industry. The author and the Australian Institute of Emergency Management worked closely with Dr Glaesser and the UNWTO to host the world's first meeting in which this integration could work, not only at the risk management level, but also in terms of in-crisis management and post-crisis recovery. The first meeting was held at the Australian Emergency Management Institute headquarters at Mount Macedon, Victoria, Australia in September 2011 and attracted tourism leaders from all key sectors of the Australian tourism industry along with tourism industry leaders from ten countries (UNWTO, 2011).

Figure 2.4 Delegates attending the first UNWTO Conference on the Integration of Emergency Management and Tourism

UNWTO Risk and Crisis Director, Dr Dirk Glaesser, is in the front row, third from the right. The author is in the front row on the far right. Keynote speaker Masatu Takamatsu, CEO Japan Tourism Marketing, is in the second row, third from the left, and Richard Gordon MBE, Director of the Centre of Disaster Management, University of Bournemouth, UK, is in the second row, first left.

Source: UNWTO, 2011. With kind permission of the UN World Tourism Organisation

One of the main speakers at that conference and its follow-up conference held in Sydney in September 2012 was the CEO of Japan Tourism Marketing Company, Mr Masatu Takamatsu. Mr Takamatsu explained that Japan's extensive experience of earthquakes and tsunamis, including the March 2011 earthquake/tsunami, had led to very close cooperation between all sectors of the Japanese tourism and event industries and emergency management providers. These involved building and siting regulations applying to hotels and resorts, evacuation procedures in the event of storm or earthquake, and the utilisation of tourism facilities to assist local civilians in shelter and transport during natural disasters. In March 2012, Mr Takamatsu organised a conference (at which the author was the keynote speaker) sponsored by the prefectural government of Okinawa, assessing Japan's progress in tourism recovery one year after the Tohoku earthquake/tsunami of March 2011, which devastated the coastal areas of Sendai Province in the North-East region of Japan's largest island, Honshu. Although Okinawa was far from the epicentre of the 2011 quake, it sought, with considerable success, to position itself as Japan's safest tourism destination through an extensive risk management programme that would protect the capital Naha from a worst-case natural disaster.

Japan, as it frequently does after a major natural disaster, enhanced its risk management protocols. These included enhanced building codes and regulations, security and evacuation procedures from hotels, resorts, event and conference centres. However, despite Japan's consistent commitment to risk management governance, there was considerable confusion that ensued after the 2011 earthquake/tsunami. This was a consequence of mixed and misleading messaging, which generated considerable distrust on the part of stakeholders of Tokyo Electric Power Company (TEPCO) and the Japan National Tourism Organisation. Poor crisis communication led to a failure to convince the media and many of Japan's tourism stakeholders that Japan was not facing a high risk of radioactivity arising from the damage caused by the earthquake/tsunami to the Fukushima nuclear power station.

In 2011, Mr Takamatsu's company, Japan Tourism Marketing, was commissioned by Okinawa's Prefectural Government to develop a tourism destination recovery and risk management master plan and strategy. Okinawa's distance (2,000 km) from the 2011 quake's epicentre meant that there was negligible physical damage and radioactivity risk from the 2011 earthquake/tsunami. However, the reputational damage to Japan generally and Okinawa specifically as a safe tourism destination following the earthquake meant that during 2011 there was a considerable decline in the number of both international and domestic tourists visiting Japan and Okinawa compared to 2010 (Takamatsu, 2014).

Takamatsu's strategy was to develop a series of risk management protocols applying to all tourism infrastructure, which would prevent damage in the event of the worst possible natural disasters. In the case of Okinawa, this was related to the most severe recorded earthquake of 1768. In parallel with these planning measures was a campaign developed in 2012–2013 to market and promote Okinawa as the world's safest destination. The master plan incorporated the outcomes of the UNWTO conference in which close relations were established by Okinawa's tourism industry with emergency management agencies to facilitate effective risk and crisis management protocols and practices. The marketing dimension of the campaign involved publication of a multilingual print and online manual and application, which advised arriving and visiting tourists of safe practices while in Okinawa. This campaign was an outstanding success and its principles implemented throughout all regions of Japan. Between 2011 and 2018, Japan enjoyed a spectacular tourism recovery from six million international tourism arrivals in 2011 to 31.2 million in 2018, representing a 500 per cent growth. This growth reflected global research, which has demonstrated that the perception and actuality of tourism safety is the principal motivating factor for destination choice today. The growth of both domestic and international tourism to Okinawa during that period significantly exceeded Japan's impressive growth.

NATIONAL AND TRANSNATIONAL STRATEGIC APPROACHES TO TOURISM CRISIS MANAGEMENT AND RECOVERY

At both global and national levels, the tourism industry has adopted collaborative national and transnational approaches to tour risk, crisis and recovery management. The UNWTO, PATA and WTTC established units within their organisations to focus on this. The Covid-19 outbreak (discussed in depth in Chapter 3), has forced all tourism industry associations to re-assess their risk, crisis and recovery strategies.

There is a realisation that managing risk, crisis and recovery in tourism can't be achieved on a sectoral-only basis. Effective crisis and recovery programmes in response to any external crisis involves tourism industry collaboration with government and emergency management agencies.

Until 2003, government travel advisories were solely determined by foreign ministries throughout the world. In 2003, Australia became the first country to invite tourism industry associations to have a consultative role in their formulation and to address issues of tourism safety for outbound travellers. This group is known as the Consular Consultative Group (CCG). The role of the CCG is to advise the Australian Department of Foreign Affairs and Trade (DFAT) on issues related to government travel advisories and global tourism safety and security applying to all Australians who travel internationally. The travel industry membership of the CCG encompasses all sectors of the tourism industry involved in outbound travel, including airlines, cruise operators, tour operators, travel writers, travel agents, consumer advocates and business travel operators and consumers. DFAT works closely with all other Australian government departments to deal with both international and global crises as they arise. The CCG has equivalent bodies with travel industry representation in the UK, Canada, US and New Zealand, all members of the 'five eyes' cooperative intelligence and security group (Beirman, 2019).

Widespread collaboration is central to the development of either a national or transnational approach to tourism risk, crisis and recovery management. This collaboration within the travel industry involves inter-sectoral collaboration. In turn, the travel industry increasingly collaborates with private sector risk management specialists, emergency management agencies and government.

National and transnational tourism organisations have developed toolkits for risk, crisis and recovery (UNWTO, 2011; PATA 2020), which emphasise the significance of collaboration within the tourism industry and between the tourism industry, emergency management, the military (on rare occasions) and government.

Discussion Questions

1. Focusing on a country of your choice, outline the key roles that government plays in tourism policy, tourism marketing and tourism risk and crisis management.
2. Discuss why some sectors of the tourism industry are able to develop strict risk and crisis management global governance, and some are unable to do so.
3. The accommodation sector is somewhat torn by its commitment to 'duty of care' for its guests and as a centre of welcoming hospitality for the destination community. How can a major accommodation provider achieve the balance between these two apparently contradictory commitments?
4. From the perspective of a tourism destination of your choice, how can you make both an effective risk management and an effective marketing tool for that destination?
5. Discuss five of the key risk management concerns that involve events and discuss the treatment of those risks with reference to a major event of your choice.

REFERENCES

American Security Project (2018) https://www.americansecurityproject.org/climate-energy-and-security/energy/todays-supply/geographic-choke-points/ Accessed 10 June 2021.

Beirman, D. (2003) *Restoring Tourism Destinations in Crisis. A Strategic Marketing Approach*. Allen & Unwin: Crows Nest, Sydney, Australia. pp. 43–68.

Beirman, D. (2014) The impact of volcanic ash clouds in 2010 and 2011 on global tourism mobility and emergency intermodal transport policy (full paper) *CAUTHE Conference 2014 Proceedings*. University of Queensland, Brisbane. pp. 71–82.

Beirman, D. (2019) 'Collaborative approaches to government travel advisories in Australia between Australia's tourism industry leadership and the Australian department of foreign affairs and trade', *Journal of Vacation Marketing*, 25(1): 71–87.

Beirman, D. and Van Walbeek, B. (2011) *Bounce Back – Tourism Risk, Crisis and Recovery Management Guide*. Pacific Asia Travel Association: Bangkok (also published in Japanese and Mandarin).

Cooper, C. (2016) *Essentials of Tourism* (2nd edition). Pearson: Harlow, UK.

Cruise Lines International Association (CLIA) (2020) https://cruising.org/en/. Accessed 2 December 2020.

International Air Transport Association (IATA) (2020) www.iata.org. Accessed 2 December 2020.

International Civil Aviation Organisation (ICAO) (2009) *Volcanic Ash Contingency Plan.* ICAO European Region: Paris. https://www.ukfsc.co.uk/wp-content/public_pdfs/Safety_Alerts/ICAO%20Volcanic%20Ash%20Contingency%20Plan%20april%202010.pdf. Assessed 2 Dec 2020.

International Civil Aviation Organisation (ICAO) (2020) www.icao.int. Accessed 2 December 2020.

International Maritime Organisation (IMO) (2020) www.imo.org. Accessed 2 December 2020.

International Maritime Organisation (IMO) (2020) *Maritime Security*. http://www.imo.org/en/OurWork/Security/Pages/MaritimeSecurity.aspx. Accessed 2 December 2020.

International SOS (2020) www.internationalsos.com. Accessed 2 December 2020.

Kenyon International Emergency Services (2020) www.kenyoninternational.com. Accessed 2 December 2020.

Pacific Asia Travel Association (2020) www.pata.org. Accessed 2 December 2020.

Safety of Life at Sea (SOLAS) (2020) *Convention of the International Maritime Organisation*. http://imo.org/en/About/Conventions/Pages/International-Convention-for-the-Safety-of-Life-at-Sea-1974.aspx. Accessed 2 December 2020.

Takamatsu, M. (2014) 'The Okinawa tourism crisis management initiatives', *International Journal of Event Management Research*, 8(1): 19–33.

The Winning Edge (2003) *Crisis: It Won't Happen to Us! Expect the unexpected and be prepared*. Pacific Asia Travel Association. Bangkok.

UN World Tourism Organisation (UNWTO) (2011) *Integration of Tourism and Emergency Management*. http://www.unwto.org/archive/asia/news/2011-10-07/integrating-tourism-and-emergency-management. Accessed 2 December 2020.

Walbeek, B. and The Winning Edge (2003) *Crisis: It Won't Happen to Us*. Pacific Asia Travel Association: Bangkok, Thailand.

3

UNPREPARED: THE GLOBAL TYRANNY OF COVID-19

INTRODUCTION

Future historians of global tourism will recall 2020 as the year of Covid-19, a previously unknown and highly contagious virus, which morphed into a global pandemic, much of it spread by international travellers (WHO, 2020). The coronavirus pandemic of 2020 triggered widespread global disruption, affecting the operation of tourism enterprises, in all sectors of the tourism and hospitality industry – in every country on earth. Core concerns discussed in this chapter are:

1. The Covid-19 pandemic's global impact on every sector of tourism called into question many of the core paradigms, dominating tourism's strategic thinking, governance, management and forecasting since 1970. Although there are prospects of tourism recovering post-pandemic, the world is likely to see significant changes in the management and operation of tourism and hospitality services. This will be most obvious for the short to medium term future while the impact of Covid-19 remains a top-of-mind consideration for prospective tourists and tourism businesses.
2. In this respect, some academics have been quick to publish papers which suggest that Covid-19 is a watershed event, which marks the end of unsustainable, exploitative mass tourism (Gossling et al., 2020; Higgins-Desbiolles, 2020).
3. A parallel theme of this chapter is that despite many previous experiences of countering pandemics, governments, medical authorities and tourism industry sectors were unprepared for the global onslaught and the consequences of Covid-19. The disease is the latest member of a family of novel corona virus outbreaks, which in recent years included: severe acute respiratory syndrome (SARS), H1N1

(Swine Flu) and Middle East Respiratory Syndrome (MERS). The lack of proactive strategies to manage a global pandemic meant that governments and the tourism industry frequently improvised responses to the many fluid challenges posed by the pandemic.

4. Primarily due to the widespread closure of international and intra-national borders, the recovery of tourism during and after the Covid-19 pandemic necessitated a staged and tightly controlled re-introduction of tourism, initially focused on local and domestic tourism. Following this would be a staged re-introduction of international tourism. This process began in mid-2020 with 'travel corridors' (also known as 'travel bubbles' or 'corona corridors'), which have operated between countries deemed by its participants to have contained the spread of the virus. The European Union (EU) led the way in implementing this strategy, although the fluidity of Covid-19's impact on destinations has meant that bubbles are easily burst when the state of Covid-19 outbreaks necessitate a reversion to lockdowns and quarantines (Beirman, 2020).

In all cases, the re-opening of intra-national and international borders required a series of re-negotiated bilateral or multilateral agreements between relevant governments. These agreements emphasised a resumption of international transport links accompanied by health-related conditions, which may apply to travel to and within participating countries (Beirman, 2020). Until an effective and readily accessible vaccine is available, there is little likelihood of tourism resuming business as usual in the short term. The process of recovery, unlike previous pandemics, which were subject to changes in consumer perceptions, is now subject to the status of regulations, determined by governments and health authorities. As a result, the timing of these changes has been gradual. In previous health-related crises, SARS (2003) and Zika (2015–2017) (Avraham and Ketter, 2016), recovery involved developing marketing campaigns focused on restoring consumer confidence based on the evidence that affected destinations were safe to revisit.

By contrast, tourism recovery post-Covid-19 involves navigating a series of government-imposed regulatory barriers. These regulations set operational limits on tourism and hospitality businesses, tourism activities and the conduct of tourists within a given destination. These precede any positive marketing campaign that the destination and its associated tourism industry may promote to stakeholders and consumers.

ENTER 2020 – A CONFIDENT TOURISM INDUSTRY WITH A TOUCH OF HUBRIS

The global tourism industry began 2020 with a justifiable sense of optimism, tinged with hubris. Optimism dominated, predicated on widespread perceptions among travel industry leaders, that tourism was resilient to the point of being indestructible. The international tourism industry expected to maintain its longstanding growth trajectory (since 2000) of four per cent per annum. The International Air Transport Association (IATA, 2020) anticipated civil aviation approaching four billion international flight sectors during 2020. Cruise Lines International Association (CLIA, 2020), the global association of the international cruise industry, planned and forecast a growth spurt from Asian markets, indicating increased enthusiasm for cruising. Generally, the global tourism industry was optimistic about growth in outbound tourism from India, South-East Asia and Latin America, which compensated for the slowing of outbound tourism growth from China since 2018. The World Travel and Tourism Council (WTTC) (the global tourism body representing global tourism's largest private companies) proudly and frequently informed stakeholders and reminded governments that tourism represented 10.3 per cent of global GDP, employed 330 million people and turned over US$8.9 trillion – an economic force with considerable clout (WTTC, 2020).

Tourism forecasters anticipated China would retain its position as the world's leading tourism generating country (145 million outbound passengers in 2019). Japan's tourism industry entered the year with a mixed sense of excitement and optimism about Tokyo's scheduled hosting of the 2020 Olympic Games in July 2020 and the benefits arising from an Olympic driven tourism boost. The major concerns of tourism planners focused on responding to rising community backlash against over-tourism in a range of popular destinations, including Venice, Barcelona, Bali, Phuket, Florence, Majorca and Machu Picchu. 'Over-tourism' was one of several sustainability concerns, arising from the rapid growth of mass tourism, since 1970 (when the Boeing 747 jumbo jet entered service and more than doubled passenger capacity on aircraft).

Political unrest (including outbreaks of violence from police and protestors) in Hong Kong, in opposition to civil restrictions imposed on the Special Autonomous Zone by China's central government, which began in mid-2019, led to a steep decline of tourism to Hong Kong – traditionally one of the world's most popular destinations. There was uncertainty surrounding the tourism implications of Brexit (the UK's economic separation from the EU). They centred on whether there would be visa-free holiday and work access between the UK and the countries of the EU. For many UK citizens who had purchased property in Europe, there were many unanswered questions about the taxation implications. Despite the problems mentioned above, during 2019, over 1.4 billion people travelled internationally (UNWTO, 2020). The key debates in global tourism associations, destination management organisations and tourism academia centred on how tourism could sustainably cope with the seemingly unending and unstoppable growth in global demand for travel.

ENTER COVID-19 – FIRST WAVE: THE VIRUS EXPOSES THE FRAGILITY OF GLOBAL TOURISM

A virus referred to as SARS-CoV-2, known as Covid-19 or widely referred as the coronavirus, was first officially recorded in the city of Wuhan, South-East China in November 2019. The virus manifests itself as a flu-like condition and for the majority of those infected it tends to behave similarly to the symptoms of a common flu. However, it has proven especially but not exclusively fatal to people with a range of pre-existing medical conditions: the elderly and frail and people with depleted immune systems. A key challenge in dealing with Covid-19 centres on the many unknowns related to this virus, some of which are mentioned below:

1. There is no absolute certainty about how the disease is transmitted from person to person, despite the fact that on 8 July 2020, WHO finally determined that it was transmitted in the air (WHO, 2020).
2. Many carriers of the virus are asymptomatic, which means they may have the virus but have neither visible or experiential symptoms.
3. There is uncertainty surrounding the incubation period of the virus after transmission. For the first few months of the outbreak, the general medical consensus was up to 14 days but this was revised to the duration of 21 days. One of the most significant challenges in dealing with Covid-19 was the fact that many carriers show no evidence of the condition visible to screening devices for a number of days.
4. There is no certainty that when a person has contracted and recovered from the disease that they are immune from future infections of Covid-19.
5. Extensive development and research by medical research teams in many parts of the world has led, by the end of 2020, to the development of several possible vaccinations to treat Covid-19. Many of these are likely to be widely available by early 2021. It is widely acknowledged that an effective vaccine would be a critical game-changer. The UK, Russia and China were the first countries to introduce mass inoculations from December 2020.
6. There is little knowledge that mutations of the virus could alter the severity of the disease and its communicability.

There is evidence that suspected cases in China were identified as early as September 2019, two months in advance of the first *officially* recorded case (Secon and Woodward, 2020; Vargas Equinas, 2020). The first officially recorded case outside China occurred in Thailand on 13 January 2020. On 30 January 2020, by which time Covid-19 outbreaks had occurred in many countries, including Thailand, Iran, Italy, France, the UK and the US, WHO declared Covid-19 to be 'a public health emergency of international concern'(WHO Interactive Timeline, 2021). This definition was clearly short of the more severe and emotionally charged word, 'pandemic'. On 11 March 2020, the United Nations WHO officially declared Covid-19 a pandemic. There has been a great deal of speculation about and criticism of WHO's apparent reluctance to declare a pandemic when it had done so for many previous communicable disease outbreaks affecting far fewer people in far fewer countries than Covid-19.

WHO's own definition of a pandemic is 'an epidemic occurring worldwide or over a wide area, crossing international boundaries and usually affecting a large number of people'. According to WHO's own definition, it was clear that Covid-19 qualified as a pandemic long before WHO's official pandemic declaration of 11 March 2020 (WHO, 2020). Covid-19's first wave was concentrated in China and then spread from China to a number of other countries.

THE SECOND AND SUBSEQUENT WAVES

By 13 December 2020, there were over 72 million recorded cases of Covid-19 globally, resulting in over 1.465 million deaths. The number of globally reported cases was increasing by well over 200,000 per day (Worldometers, 2021). By 30 November 2020, America represented almost half of global cases. The US had the world's highest number of Covid-19 cases and deaths, recording 16.53 million cases and 305,000 deaths while India, ranking second, had recorded 9.85 million cases and 143,000 deaths (WHO, 2020). By this time, China ranked 79th in the world for the official number of cases and the epicentre of the pandemic had clearly shifted from East Asia to the Americas, Europe and the Indian subcontinent. Brazil had had the unwanted distinction of the world's third ranking country for Covid-19 cases, recording 6.88 million cases and 181,000 deaths. Towards the end of 2020 major spikes in the number of cases and deaths occurred in Europe and the UK. By 13 December 2020, the UK registered 1.83 million cases and just over 64,000 deaths (the equal highest death rate in Europe with Italy). The impact of Covid-19 has affected all social classes and all age groups. During 2020, three heads of government were among the millions who contracted the disease, UK Prime Minister Boris Johnson, Brazilian President Jair Bolsonaro and US President, Donald Trump. In the UK, the heirs to the throne, Prince Charles and his son Prince William, also contracted Covid-19.

Apart from Antarctica, a few Pacific island nations, North Korea and Turkmenistan, cases of Covid-19 were recorded in almost every country on Earth. The Covid-19 pandemic, caused by a biological entity (biophage, better known as a virus), would result in a scale of devastating impacts on tourism never experienced during any previous crisis arising from any other cause. (This book seeks to thematically address the causation of most tourism crises.) Governmental responses to Covid-19, at national, provincial/state/regional and local level, involved a combination of border closures and lockdowns imposed on vulnerable communities and areas suspected of being virus 'hotspots'. Transport bans and restrictions, event closures, restrictive social distancing regulations and strict hygiene measures were applied to citizens, businesses and tourists by all three tiers of governments worldwide.

These regulations varied from country to country and even regional or local jurisdictions within countries, manifesting themselves in limitations or bans on air travel, cruises, rail and bus travel, events, conferences and conventions, sporting fixtures, theatres, cinemas, restaurants, shops, markets, cafes, pubs and visitor access to attractions. For the first time since the world wars, the Olympic Games, scheduled for Tokyo in July 2020, was postponed until July 2021. The postponement of the 2020 Olympics was the most prominent of thousands of cancelled or postponed sporting, cultural events and festivals during 2020.

The second wave of Covid-19 essentially signalled that the pandemic had morphed from a China-centric health threat to a global problem. From March 2020, many cases of Covid-19 emerged in localities with no discernible connection with China.

Figure 3.1 The global impact of Covid-19 on global tourism numbers, Jan–April 2020

As discussed during this chapter, Covid-19 established itself as the greatest disruptor of global tourism in recent history, with the potential to be a paradigm shifter for the global tourism and hospitality industry. In that sense, Covid-19 is far more than just another pandemic (discussed in Chapter 7) and the scale of its impact on tourism justifies a separate chapter in this book. As Figure 3.1 illustrates, the impact on global tourism movements during the first four months of 2020 was severe, even accounting for the fact that January 2020 was unaffected.

While this book focuses on tourism, it would be remiss not to acknowledge that Covid-19 severely disrupted every sector of international trade, commerce, agriculture, industry, education and international political, social, scientific, technological, sporting and cultural exchanges. However, the evidence suggests that the global tourism industry is among if not *the* most severely impacted sector of the global economy. For those countries (including many developing countries) with a high level of economic reliance on tourism, the economic and social impacts is potentially catastrophic. By November 2020, it was estimated by the World Travel and Tourism Council (WTTC) that 190 million jobs in tourism and hospitality had been lost and the financial losses to tourism enterprises was at least US$2 trillion (WTTC, 2020). By any measure, Covid-19 had severely undermined the validity of tourism's fundamental paradigm of its existential resilience.

Pandemics, travel and tourism have endured a toxic association (a *danse macabre*) for centuries. The movement of people across international boundaries through travel played an integral role in spreading a vast number of communicable diseases all over the world. Throughout history, human travel movements have resulted in communicable diseases spread with occasionally 'genocidal' consequences. Entire civilizations including the Incas, Tasmanian Aborigines and Mayans were annihilated following exposure to diseases transmitted by foreign invaders who were immune from diseases such as measles or influenza for which the indigenous inhabitants lacked natural antibody resistance, and medicinal cures were either unavailable or yet to be invented.

From 2000 to 2019 the global tourism industry responded to several pandemic outbreaks, including: SARS (2003), H1N1 (swine flu 2009–2010), avian flu (2014) and Ebola in West Africa (2015), Zika (2016–2017) and MERS (2018). In most instances, except for H1N1 the spread of the diseases were regional rather than global. The most extensive pandemic outbreak in recent years (until the outbreak of Covid-19) was H1N1 (swine flu outbreak). According to the Centers for Disease Control and Prevention (CDC) (CDC, 2020), H1N1 infected 65 million to 1.2 billion people globally between March 2009 and September 2010 and caused between 270,000 and 570,000 deaths. The uncertain mortality rate was due to many fatalities claimed as attributable to H1N1 but involved people who concurrently suffered from other illnesses. At worst, the mortality rate of H1N1 was 0.03 per cent. This is discussed in Chapter 7.

In late 2009, a vaccine that effectively protected 60 per cent of H1N1 patients was introduced. It is notable that some of the diseases mentioned above had a far higher mortality rate than the two per cent attributed to Covid-19. This was certainly the case with Ebola, MERS and SARS. However, the number of people afflicted by these viral conditions was small compared to Covid-19. Despite the fear generated by SARS in Asia and Canada during 2003, there were only 8,500 cases recorded, resulting in 800 deaths. MERS had a very high mortality rate of 30 per cent but was not easily transmissible and in 2017–2018 there were fewer than 2,000 cases worldwide. It is also an uncomfortable fact that no curative or preventative vaccines existed for SARS, Ebola or MERS. Despite intensive efforts from thousands of medical researchers worldwide, the search for a vaccine to counter Covid-19 led to a range of vaccinations being introduced by the end of 2020.

A unique characteristic of the Covid-19 pandemic was the series of government led measures taken to isolate and quarantine infected people and limit the spread of the virus throughout the world. In the case of most recent pandemic outbreaks, people have been screened at entry ports and gateway airports and if found to be infected, subsequently quarantined. In the case of Ebola (2015), in West Africa, foreigner entry and outbound tourism was heavily restricted to the four most affected countries, Liberia, Guinea, Gambia and Sierra Leone. However, the impact on global tourism was minute as these countries were not tourism hotspots. By contrast, the Covid-19 pandemic has led to all countries (including countries with few or no cases) imposing a variety of regulatory barriers – many of which excluded or restricted tourism visitation to prevent the spread of the pandemic.

Many of these regulatory measures to contain Covid-19, including border closures, area lockdowns, enhanced hygiene practices and social distancing, have proven to be inimical to the traditional operation of tourism and hospitality enterprises. Restrictions imposed to limit the free flow of people between international and domestic destinations, ran counter to the fundamental *raison d'être* of tourism. During the Covid-19 outbreak, border and jurisdictional barriers restricting cross-border travel was applied to travellers by all three tiers of government worldwide.

The immediate impact of the border restrictions was a massive reduction in the number and frequency of international and domestic airline services – a trend evident from March 2020. As China was the first country to experience a significant outbreak of Covid-19, the US was the first of many countries (January 2020) to officially ban the entry of travellers from China. However, by the time those bans were enforced, a small minority of Chinese travellers had unwittingly spread Covid-19 to the many countries they had visited prior to mid-January 2020. In January 2020, the Chinese government locked down movement throughout the city of Wuhan (population 11 million) and the surrounding Hubei province from 23 January to 8 April. This region was identified as the epicentre of the initial Covid-19 outbreak. To reinforce the Wuhan lockdown,

China banned outbound international travel for its citizens, from late January 2020 – just ahead of the Chinese New Year holidays (the traditional peak period for Chinese international and domestic travel). This move instantly placed a travel freeze on ten per cent of the global tourism market.

As Covid-19 rapidly expanded into a global pandemic, most countries in the world imposed bans or strict limitations on international tourism entry. In many countries, foreign visitors and nationals arriving from countries deemed at risk were required to undergo a period of quarantine – usually but not exclusively for 14 days. From 5 February 2020, Covid-19 outbreaks on three cruise ships docked in Asian ports (*Diamond Princess*, Yokohama, Japan; *Westerdam*, Sihanoukville, Cambodia; and *World Dream*, Hong Kong) led cruise ships to become a focus of global concern. Quarantine regulations on cruise ships, their passengers and crews were implemented by Japanese, Cambodian and Hong Kong port authorities, when cruise ships were identified as posing significant dangers as incubators and vulnerable environments for the spread of Covid-19. Over 700 cases were identified on the *Diamond Princess* alone (WHO, 2020). In the Australian state of New South Wales (NSW), the disembarkation of 2,600 passengers in Sydney (approved by the NSW Port Authority) from the Carnival Cruise ship, *Ruby Princess* on 19 March 2020 led to the largest single source of Covid-19 cases in NSW, involving over 200 cases and 19 deaths attributed to passengers from that ship. This incident was subject to a NSW government (2020) Special Commission of enquiry triggering an intense debate (much of it played out in the Australian media) over determining the extent to which fault lay with the cruise company (Carnival Cruising), or the NSW Port Authority and the NSW state government. The findings released in August 2020 held the NSW government as primarily responsible. From late March 2020, cruise ship companies largely suspended operations until late 2020, as few ports were prepared to accept them. In September 2020, CLIA released a range of upgraded health and safety protocols to apply to global cruising and hasten its worldwide resumption (CLIA, 2020).

The global cruise sector, governed globally by the UN International Maritime Organisation (IMO), has an extensive range of risk management protocols and ship-based responsibilities under its Saving of Life at Sea (SOLAS) charter initiated in 1914. However, the issue of managing communicable diseases (plague ships) has long been a grey area. Cruise ship operators are authorised and legally obliged to deal with health-related issues on the high seas. They are required to advise local authorities of disease outbreaks when in territorial waters. However, when a ship is sailing within or docking in territorial waters, the governing jurisdiction assumes responsibility for the management, and if necessary, isolation and quarantine of cruise ship passengers and crew. As such, the port authority determines whether passengers need to remain onboard the ship for quarantine purposes or determine the safety of passengers and crew to disembark. Ship's crew and their medical officers are responsible to accurately report medical emergencies or onboard epidemics to those authorities. Responses to those reports are the responsibility of the port's governing authority.

By May 2020, practically every country on Earth had imposed restrictions on travel movements irrespective of the extent of their exposure to the virus. The restrictions on tourism transport movements were matched by social distance regulations, which set the number of people who could dine at cafes and restaurants, visit places of entertainments, pubs and wine bars, indoor attractions such as museums and art galleries, sporting events, meetings, conventions and exhibitions, schools, colleges and shopping malls. Social distance regulations limited the number of people who could visit private homes, attend weddings, funerals, religious services and any other gathering. It also limited participant numbers of group tours. In many cases, the restrictions (especially when certain places imposed a lockdown) were so severe that many of these businesses ceased trading. Thousands of events, meetings and conventions were cancelled, postponed or (when practical) conducted online.

Initially (January 2020), China was the country with the highest number of cases and deaths but by late November 2020, the US emerged as the country with both the highest case numbers (13.75 million) and the highest number of deaths (273,000). As discussed earlier, by 30 November 2020, China was ranked 71st in the world for its 86,512 number of recorded cases, of which 4,634 resulted in deaths. China's ten-week lockdown of Wuhan/Hubei between January–April 2020 led to a significant reduction of subsequent Covid-19 case numbers recorded in China.

The above figures are those supplied by government health authorities to WHO. There is a strong likelihood that the actual figures (in many cases due to the lack of testing in many countries) may be considerably higher.

THE GLOBAL POLITICAL AND ECONOMIC DIMENSION

The global scale of the Covid-19 crisis involved a significant political and economic dimension. Although the political and economic implications of the pandemic are not the focus of this chapter, they can't be ignored as they had a direct impact on tourism. There was a link between the overall tension and rivalry, which has existed between the US and China during the past decade, and US government accusations against China for allegedly covering up the emergence of Covid-19, strenuously denied by China's government. The Wuhan lockdown and the ban China imposed on outbound international travel in January 2020 was cited by the Chinese government to support its claim of adopting the world's most far-reaching policies to stem the spread of Covid-19. The USA was the first (of many countries) to ban tourism from China, a measure that lost its diplomatic impact when China virtually banned outbound tourism.

In February 2020, the Trump administration announced the cessation of US government funding of WHO on the grounds of WHO's alleged reluctance to declare a pandemic early in 2020. President Donald Trump accused WHO of 'pandering to China's sensibilities. As the USA annually contributed 20 per cent of WHO's annual budget, this represented a significant blow to WHO's global operation. Microsoft CEO, Bill Gates, one of the world's wealthiest men, pledged to contribute financial support to WHO, equivalent to that of the US government. Nevertheless, elements of the US government's criticism of WHO, especially China's resistance to an independent and transparent investigation, was shared by a number of national governments, including the UK, Australia, Canada, France and Italy.

National governments played a prominent role in introducing a range of fiscal measures and social security funding to cushion their citizens from the severe financial and employment impacts of Covid-19 related regulations on a wide range of small and large businesses, including tourism. Although the economic protective measures have varied between countries, governments, irrespective of political ideology, have provided bailouts, tax holidays, loan repayment and rent deferrals for both strategically important industries (including airlines) and businesses where operations have been affected by lockdowns and social distancing regulations.

Some governments have also spent large amounts of public money to support businesses and their employees to protect jobs for businesses required to operate under restricted trading and operational conditions. Many governments increased social services payments and even paid for the hotel accommodation of citizens who returned from overseas and were required to undergo a period of quarantine. A common concern of all governments in the world has focused on the concern that prolonged economic stagnation, because of the pandemic, could lead to a global recession or depression.

There was widespread global suspicion that China, despite the lockdown of Wuhan and other regions affected by Covid-19, had covered up the early stages of the viral outbreak. Australia's government was among the most vocal national governments calling on the UN to investigate the early stages of the outbreak in China. China's initial response to Australia's call was one of outrage. The Chinese government-imposed tariffs on a number of Australian export products. Much to the Chinese government's chagrin, Australia was not alone in demanding a comprehensive and transparent enquiry and a World Health Assembly (2020) (the policy-making body for the UN's WHO) vote called for such an enquiry on 19 May 2020 (ironically and ultimately co-sponsored by China) passed by unanimous votes of all 134 delegates.

While most border restrictions imposed since the viral outbreak sought to limit the spread of Covid-19, some border closures involved a political dimension. Apart from the US bans on Chinese visitation, many countries have been selective in applying visitor entry to certain nationalities. Due to the high incidence of Covid-19 in the US,

Brazil and Mexico, EU countries excluded US, Brazilian and Mexican citizens from entry to their partially re-opened borders in July 2020.

Health-based restrictions have sometimes served as a convenient cover for politically based motives to restrict the entry of nationalities politically out of favour with host countries.

In Europe, moves towards the finalisation of the UK's formal economic separation from the EU (better known as Brexit) was delayed by the practical difficulty in arranging face-to-face discussions with relevant leaders. Major international diplomatic discussions and exchanges have employed online meetings and conferences replacing face-to-face meetings as a means to minimise exposure to the pandemic.

The Covid-19 outbreak engendered a range of conspiracy theories, many of them communicated via social media and websites. Iran's government, much of the Arab and Islamic world and extremists fringe groups from both the left and right wings of politics have blamed Zionists, Israel, the Jews or all three for Covid-19 – alongside their arch-nemesis ('Great Satan'), the US. Historically, accusations against Jews instigating or spreading pandemics have existed for centuries, going back to the bubonic plague (Black Death) outbreak of the 14th century. Conspiracy sites revealed a wide range of suspects for the manufacture and spread of Covid-19, including the Chinese government, American government, Bill Gates, Canadians, manufacturers of the 5G mobile phone network, the Illuminati, among a rogue's gallery of other suspects. Much of the speculation among conspiracy theorists and serious virology researchers focuses on the origins of the Covid-19 outbreak in Wuhan. There is considerable debate over the first cases in Wuhan, speculating whether the disease originated from wet markets in Wuhan or an infectious disease laboratory near Wuhan. Conspiracy theories, harbouring alternative views to mainstream opinions, need not be summarily dismissed. However, their proponents must be subject to the same rigour as scholars and scientists by providing compelling factual evidence to back their claims (Braun, 2020; Nirenstein, 2020; Oriel, 2020).

COVID-19 PANDEMIC AS A PARADIGM SHIFTER IN GLOBAL TOURISM

Since the end of World War II in 1945, the global tourism industry operated on a range of common assumptions, which coloured its pre-Covid-19 paradigm (worldview or *weltanschauung*).

The core paradigm of global tourism (pre-Covid-19) asserted that tourism is a positive force in the world, a growing sector of global economic and social activity committed to the free movement of people and peaceful cultural exchange, which if managed sustainably, can potentially address many of the world's environmental, social and political problems, and is fundamentally resilient.

Central to the above paradigm were benefits that tourism derived from the increasingly free movement of people across international borders. Prior to the Covid-19 pandemic, relatively unfettered movement of short-stay tourists, across international borders (as opposed to other categories of arrivals), was facilitated by most national governments. Measures included easing arrival processes, procedures, cost and barriers for foreign tourists to cross borders, in most cases, aided by the easing of visa restrictions and enabling tourists to complete visa documentation online from their country of departure. The EU introduced the *Schengen* visa, which enabled tourist access to most EU countries for selected nationalities on arrival in a participating EU point of entry. EU citizens were able to cross national frontiers in most other EU countries with virtually no formalities (the UK was a notable exception to this even before Brexit). However, the easing of visa entry requirements and entry restrictions for the nationals of most countries has been far from universal. The state of international political relations between nations and security concerns has led to a gradation of visa rules and restrictions between certain countries.

Advances in transport technology, especially since the introduction of the high-capacity Boeing 747 jumbo jet in 1970, including long-range flights, large capacity airlines, massive cruise ships, the development and expansion of national and international high-speed rail links, and the expansion of highway networks for motor vehicles, resulted in a significant reduction of travel times between domestic and international destinations. The massive growth in passenger capacity for airlines, cruise ships and trains since 1970 enabled a significant reduction of the unit cost of travel for consumers both in real dollar terms and actual dollars spent.

The increasing affordability of travel led to significant increase in patronage. By 2019, the world's largest passenger jet aircraft (the A380) could comfortably accommodate more than 500 passengers (in economy class seating configuration). The world's largest cruise ship, Royal Caribbean's *Symphony of the Seas* carried 6,680 passengers and 2,200 crew (Royal Caribbean, 2020). The high passenger capacities of planes, ships, trains and hotels enabled economies of scale, making travel increasingly affordable to a wider demographic of people. On land, the growth in capacity applied to hotels, resorts and a growing range of accommodation providers. In many popular tourism destinations, hotels and resorts could count rooms in the thousands with room configurations to match a variety of guest budgets. Accommodation options have expanded in line with the growth of market demand. At many popular tourism destinations, visitors could choose from a wide range of accommodation options. These included: luxury hotels, motels, capsule hotels, resorts, eco lodges, backpacker lodges, youth hostels, caravan and camping parks, traditional bed and breakfast establishments, accommodation representing authentic national accommodation in specific countries, pub stays and home stays dominated by the rapid growth of Air BnB, which offers more accommodation space than any hotel chain in the world. In short, the growth of mass travel had also demographically, *democratised tourism*. This has meant that between 1970 and until the beginning of 2020 tourism became increasingly affordable to a wider cross section of people than ever before.

To illustrate how this has worked in practice, in 1981, a low-cost return airfare between Sydney and London cost about US$2,000. By 2019, a low-cost return airfare between Sydney and London cost US$1,000. Reductions in the cost of long-haul air fares was accompanied by a similar reduction to short and medium haul air fares applying to popular and competitive airline routes. Economies of scale effectively reduced the cost of both international and domestic air fares, international and regional rail fares, point-to-point coach tickets, cruise tickets, hotel rooms, resort stays, and in many cases, package tours. The introduction of low-cost carriers during the 1970s and their global proliferation since 2000 has contributed to the reduction of point-to-point airline fares applying to most major tourism destinations. Coupled with this, the rise of the middle class in many Asian countries since 2010 led to international travel becoming increasingly affordable to a growing number of people, worldwide.

These developments in global tourism led to an overwhelmingly optimistic paradigm that (apart from some disruptive events and circumstances) global tourism would continue to grow indefinitely. However, while tourism would need to make allowances for sustainability, safety and ethical business practice, it would continue its onward march to become the world's largest industry and the global economy's greatest employer.

The pre-Covid-19 paradigm for tourism centred on a number of key assumptions:

- Tourism would continue to grow (apart from some short-term disruptions and downturns) indefinitely.
- Tourism would become increasingly democratic and affordable to a larger number of people in both the developed world and the developing world.
- Tourism and hospitality enterprises could successfully operate as high volume and low yield enterprises. Apart from a few business collapses, most tourism businesses would remain viable provided they generated sufficient volume.
- Tourism was viewed as a desirable industry by governments all over the world as a means to advance national economies, infrastructure, influence regional development, create employment in cities and regional areas, and positively showcase destination countries and regions within the country.
- In order to maximise their economic benefit from tourism, countries around the world would, subject to security considerations, facilitate international tourism and minimise entry restrictions for international tourism arrivals.
- The world was shrinking and tourism was a means to bring the world closer together.
- Travel and events were essential parts of conducting business and education.

- Although the tourism industry accepted the likelihood that crisis events could temporarily disrupt tourism, consumer demand for tourism and the resilience of the travel industry would enable rapid recovery of tourism globally and in affected destinations.
- Risk management protocols applying to all sectors of tourism had effectively minimised the likelihood on long-term disruption to tourism.
- Travel agents and online retailers believed that if one destination or a destination region became problematic to promote or sell, there were many alternative destinations.
- Tourism would be an increasingly popular area for vocational and higher education.
- The growth of tourism and hospitality offered a wide range of employment opportunities for the full range of skills and managerial levels.
- Tourism provided an opportunity to promote and support unique cultural characteristics of a destination and its people.
- Meetings, conventions, exhibitions and events were a magnet for international and domestic visitation.
- Airline, sea and land transport facilities and infrastructure access is necessary to facilitate access for international and domestic tourists.
- The dispersion of tourists within a destination region was a major asset to ensure visitor stays were of optimal duration and earned the destination greater and more dispersed income.

Covid-19's disruptive influence on global tourism centred on a number of key factors:

1. *The spatial factor*

   Most recent pandemic events, including SARS, Ebola, MERS and Zika had an impact on specific regions of the world or on specific groups of people. Of recent pandemics, only H1N1 had a global reach and this was ameliorated by a very low mortality rate and a vaccine that treated most at-risk people. By contrast, Covid-19 rapidly spread throughout most of the world, continued to intensify, has only been subject to medical intervention from late 2020 and has a relatively high mortality rate of just over two per cent. From a tourism perspective, one of the main regulatory responses to Covid-19, widespread closures of international and intra-national borders, limited the international and (in some countries) intra-national freedom of tourism movement.

   The most visible outcome of border closures was the widespread cancellation and route closures of international, and in some countries, domestic airline services. As government regulations, rather than consumer sentiment, determined and regulated these closures, they tended to prolong regulatory changes. All bilateral agreements required government–government negotiations, including agreements to reinstate travel and transport links. Unlike previous pandemic events, Covid-19 had a high, global severity impact. Traditional pre-Covid-19 destinations and tourism enterprise resilience and recovery strategies as they applied to past pandemics were inapplicable, due to the tourism industry's lack of control on government-imposed regulatory changes and recovery messages.

2. *The temporal factor*

In the pre-Covid-19 tourism paradigm, the travel industry had a high degree of control on the timing of a crisis and recovery. The primary role of government in responding to Covid-19 effectively removed the tourism industry from any control over the timing of crisis management and the setting of a recovery agenda.

Although the first officially confirmed cases of Covid-19 occurred in late 2019, in China, the pandemic rapidly encompassed most countries by March 2020 and its spread accelerated during 2020. As the pandemic intensified by over 200,000 new cases per day and in the absence of a preventative and a treatment (vaccine) until the end of 2020, the pandemic dominated medical and tourism industry concerns during 2020 and well beyond that. The introduction and distribution of a viable vaccine requires considerable time before the pandemic is controlled. As the majority of restrictions on tourism activity (border restrictions and social distancing regulations) are subject to government regulations, rescinding or easing them is likely to be gradual rather than instant.

The strategic approach towards recovery of both domestic and international tourism is likely to operate in a series of stages, as opposed to a blanket (return to 2019) resumption of tourism. A term much used in 2020 was the introduction of 'travel bubbles' or 'travel corridors'. Travel bubbles/corridors applied to selected international routes in which the participating governments had agreed that travel could operate under stringent conditions. They applied to selected nationalities. The first evidence of travel bubbles/corridors operating during the pandemic appeared at the end of June 2020 in a number of EU countries (Beirman, 2020; Thiessen, 2020). However, the fluidity of the Covid-19 pandemic has meant that the establishment of a travel corridor was no guarantee that it will remain in place. In early July 2020, British travellers, were permitted to visit Spain and return without having to self-quarantine. By the end of July 2020, a spike in the number of Covid-19 cases in Spain resulted in the British government changing the rules and re-applying a quarantine requirement for British travellers returning from Spain.

The concept of travel bubbles/corridors was widely regarded as an initial test case towards a more widespread opening of international travel routes for those countries in which Covid-19 was perceived as being contained. Many of the most enthusiastic advocates of travel bubbles/corridors were governments of those countries in which tourism was a significant element in their national economy. However, the fluidity of the pandemic means that no agreements were binding.

3. *The risk, crisis and recovery management factor – sectoral perspectives*

Two of the key words in the lexicon of risk management are *likelihood* and *consequence*. The likelihood of a pandemic on the scale of Covid-19 was always high. Pandemics in one form or another occurred at regular intervals. The last global pandemic prior to Covid-19 was Zika in 2016–2017, but as its morbidity impact was confined to newborn children; the core response was to ensure that pregnant women avoided affected areas. H1N1 in 2009–2010 affected 1.4 billion people, but H1N1's combination of a low mortality rate (0.3 per cent), reduced further by a vaccine with a 60 per cent success rate, resulted in H1N1 being perceived as more of a health hazard than a crisis.

Some transmissible diseases with a high mortality rate, including AIDS, Ebola and MERS, proved to be high-risk infections for some segments of the population and specific geographical regions. While the risk outside those vulnerable groups or regions is limited, the 2014–2016 Ebola outbreak was quarantined and restricted to four West African countries.

Since the beginning of the 21st century, the tourism industry in cooperation with emergency agencies developed strategies employed in the event of a pandemic. They included medical screening of passengers at arrival and departure at air and seaports, and occasionally, international railway stations. These were widely employed during the SARS outbreak of 2003 and H1N1 in 2009–2010. Other measures included isolation and quarantine of people with symptoms, undertaking medical testing ahead of a package tour by tour operators, enhanced hygiene procedures in gatherings for events, entertainment venues or visits to attractions.

Until Covid-19 became a well-established threat, most tourism destinations, businesses and governments actually did little or nothing to address or prepare for the threat. Airport screening of passengers for Covid-19 during January and February 2020 was the exception rather than the rule.

In fact, it was not until 20 March 2020 that Airports Council International (the global association of airports) released guidelines for Covid-19 for its member airports (ACI, 2020).

From a broader perspective, governments and health authorities in many countries were slow to respond to the threat of Covid-19 and some governments miscalculated the gravity of the pandemic's threat. The global tourism industry was similarly unprepared for Covid-19 and sluggish to respond to it as a crisis. Many tourism products and services operate on the basis that they are perishable products. Unsold airline seats, train seats, coach seats, hotel space, tour places, dining tables etc. become a liability and the drying up of tourism was literally eroding the viability of most tourism businesses. Each sector of the tourism industry faced its own set of risk management challenges arising from the pandemic.

Airlines

The airline sector was severely affected by the Covid-19 outbreak. According to IATA (2020), the global downturn in air traffic exceeded 90 per cent since the widespread closure of international and many intra-national borders took effect in late March 2020. This has led to literally thousands of airline routes rendered inoperable for an indeterminate period. Airlines, in common with other sectors of the tourism industry, operated a business model based on high volume, high passenger loads and extremely thin yield margins per passenger. They are an illustrative case of perishable business operations.

A long-haul economy airfare between Sydney and London generally costs the consumer about USD$1,500. The airline is usually lucky to yield $10 in profit per passenger after all the airline's costs are included (Elsegood, 2019). Yields are greater for business and first-class passengers. 2018 and 2019 were relatively profitable years for most international carriers primarily because most of them were operating with full passenger loads. Low profit margins left airlines with very little in the way of financial reserves in the event of a business downturn. Covid-19 represented a downturn beyond the worst nightmares of airline executives. Many were quick request bailouts from governments. For government-owned airlines and those carriers perceived as a national strategic asset, government support was readily maintained.

As Figure 3.2 illustrates, the decline in airline business was both instant and precipitous. The core question for airlines was: How long will it take until airline business recovers to 2019 levels? IATA's own forecasting allows for variable projections but the general consensus from IATA members is that recovery to 2019 levels may not occur until 2024 (IATA, 2020).

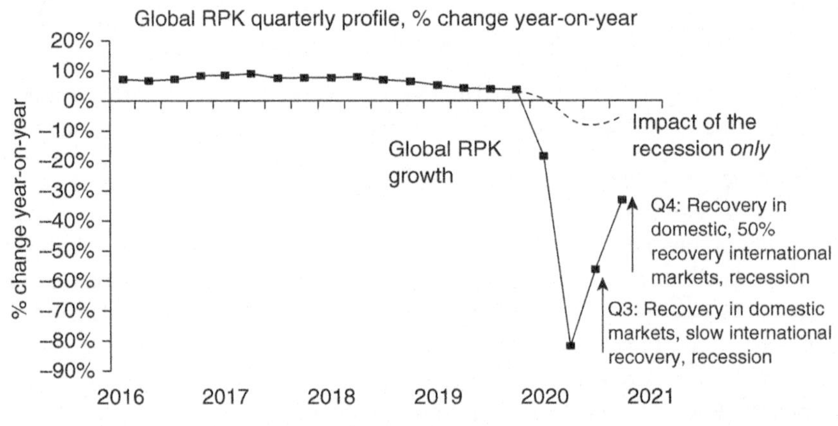

Figure 3.2 The impact of Covid-19 on global airline revenue per kilometre

Source: IATA, 2020. Used with kind permission of IATA International Air Transport Association.

Airlines were reluctant to introduce any visible measure to facilitate social distancing although they did enhance on-board hygiene measures. IATA mentioned that the measures it had introduced during the SARS outbreak to filter ventilation systems was sufficient to prevent Covid-19. Unfortunately, no air conditioning filtration can prevent the transmission of Covid-19 by people in close physical contact with each other. More recently, IATA supported passengers and crew being required to wear surgical masks on most international air routes. However, the lack of passengers on many international air routes (often loadings imposed by governments) assured social distancing by default. Weaning airlines away from high passenger loads to 'pay the bills' was not a measure that airlines voluntarily initiated.

Land Transport

Trains, coaches and even rental cars have been subject to social distance restrictions. Many operators quickly adapted. However, in common with airlines, those routes involving border crossings significantly complicated the operation of international and inter-state/provincial/prefectural land transport. Rail (due to the far greater passenger space they offer compared to aircraft) may derive some benefit from passengers concerned about maintaining safe social distancing while travelling during the Covid-19 pandemic.

Cruising

The cruise sector was a significant sector of the tourism industry, relatively unprepared for the impact of Covid-19. Although the cruise industry is well aware of risks applying to communicable diseases, the recent growth in size and passenger capacity of cruise ships included the risk that they were a floating pandemic disaster waiting to happen. Cruising is a very sociable environment for the thousands of passengers and crew who sail for leisure, pleasure and uninhibited behaviour. Most meals served on cruise ships were served in the form of buffets (which encompass a range of hygiene and social distancing problems), staged entertainment, which required close contact in crowds, and a wide range of social activities, which were inimical to social distancing.

Social distance regulations that were progressively imposed by government in many countries run counter to the onboard ambience of a cruise experience. All major cruise operators have trained medical staff, clinics, small infirmaries and isolation procedures onboard. However, these are designed to cope for a few ill passengers rather than an outbreak, which could potentially infect over 6,000 passengers and 2,000 crew, on large cruise ships. In September 2020, CLIA released an upgraded set of health and safety protocols for global cruise ship operators in response to Covid-19 related concerns (CLIA Global, 2020) in the hope that this would accelerate a widespread return to cruising.

Meetings, Incentives, Conventions/Conferences and Exhibitions/Events

The meeting, incentives, conventions/conferences and exhibitions (MICE) sector of tourism was ill-prepared and largely ill-equipped to cope with the risk implications of Covid-19. In all cases, MICE businesses centred on hosting large numbers of people at close quarters. In most instances, there was a strong emphasis on social interaction for the non-business aspects of MICE events, including food and beverage catering. Social distancing regulations and restrictions were simply toxic to the viability of the MICE market.

Some meetings and conferences were able to be conducted remotely using online technologies but while this was a useful alternative means to share content and information, it lacked the ambience, experiential, networking and commercial potential of face-to-face events. A significant challenge for event organisers involved setting price points for virtual conferences and events. The MICE market is traditionally labour intensive. The lack of face-to-face events had an immediate and severe impact on job losses in the sector. The real winners of the Covid-19 outbreak were online meeting sites such as Zoom, Microsoft Teams, Webex and GoToMeeting.

As mentioned earlier in this chapter, thousands of business, cultural and sporting events were cancelled or postponed. This was due to government-imposed Covid-19 related restrictions. Many event organisers were unprepared to adhere to the altered regulatory environment or adapt to online technologies.

Tourism Training and Education

Schools, universities, colleges and private training institutions were required to adapt to online teaching and learning to maintain course and subject commitments and schedules. In the case of most educational institutions (including universities), this change was achieved in a matter of days, due largely to the fact that the technology was well established. However, the most significant impact of Covid-19 on many educational institutions, especially universities, were limitations set by governments on the movement of international students.

Accommodation and Hospitality

The accommodation/hospitality sectors of the tourism industry were among the most heavily impacted sectors of global tourism. There were impacts arising from Covid-19 for which the sectors were clearly unprepared; this especially applied to social distancing restrictions on in-house events, health facilities and in-house dining. Those accommodation providers typically dependent on cross-border, international or intranational tourism were bereft of guests if they were located in a jurisdiction closed to key source markets. Social distancing regulations restricted the operation and guest capacity of accommodation providers, restaurants, cafes, pubs, clubs, theatres, cinemas, places of entertainment, function centres, etc.

Some accommodation providers, notably hotels, derived some turnover from governments hiring them as quarantine centres for travellers who were required to undergo mandatory quarantine when arriving from international or inter-state/provincial/prefectural travel. Small hotels, motels and private accommodation such as home stays were able to operate normally (with enhanced hygiene and social distancing measures), but the reduction in the number of travellers has affected the number of guests in most sectors of the accommodation industry. Notably, some Air BnB properties experienced a surge in local area bookings as many people sought alternative 'escape' accommodation from home-based lockdowns. Hotels, which relied on income from in-property dining, banqueting and functions, found their operations were severely restricted by social distancing regulations. Traditional buffet meals (especially breakfasts) offered by many accommodation providers were banned by many regulators.

Some restaurants and cafes were able to provide a skeleton service focused on providing take-away food and beverages and were able to re-open on a limited basis when and if social distancing regulations were eased. Shopping malls, retail outlets and markets were required to observe social distance regulations.

Attractions

Demand for attractions was affected by the reduction in visitation from outside of the destination regions and by social distancing regulations. This especially but not exclusively applied to ski resorts, beaches and other seasonal attractions. The financial

Figure 3.3 Bondi Beach in Australia, March 2020, closed due to Covid-19 concerns

Photo reprinted with kind permission and courtesy of Ms Darryl Bourke.

viability of many (commercial) attractions requires an optimum number of paying visitors. Many attractions were required to limit the number of attendees and regulate the physical spread of visitors (especially in indoor or enclosed attractions). Not all attractions suffered a loss of business due to Covid-19. Some outdoor attractions (notably outdoor national parks) experienced a surge of visitation largely because they remained among the few attractions less restricted by regulations.

Tour Operators

The essence of a tour operator's work involves collaboration with a number of suppliers and operators to package tour programmes. Risk management is widely regarded as an essential part of a tour operator's management toolkit. The author worked closely with the Council of Australian Tour Operators (CATO) to develop a risk and crisis management manual in 2016, which was revised in 2019 (Beirman, 2016 and 2019).

In those years, CATO ran extensive workshops involving a number of presenters and examined the response to a range of risks, including pandemics. Globally, both inbound and outbound tour operators are exceedingly conscious of risk. The Travel Corporation, the world's largest multi-branded company of tour operators, has been active in training its management and staff on risk and crisis management procedures. Yet, conscious as they have been about risk management, tour operators were unprepared for the scale and impact of the Covid-19 outbreak on their businesses.

Tour operators, whether local, inbound or outbound domestic and international, were affected by Covid-19 on two fronts. The first front was the limitation on cross-border tourist movements, preventing tourists from travelling to tourism destination regions to join tours. This led to an avalanche in tour cancellations. An issue related to this front was that suddenly a vast portion of tour operator product (especially international product) became unmarketable for an indeterminate time. The second front involved social distancing regulations, which affected the size of tour groups, transporting them to attractions and enabling entry into attractions. Although some tour operators have marketed small groups or private tours for many years, most tour operators, especially those who operated large group and coach tours, were affected by one or both of the above fronts. The majority of tour operators tended to offer increasingly flexible tour conditions, enabling travellers to postpone package tours affected by the pandemic. Refunds were often complex because many tour operators worked with multiple principals whose booking and cancellation conditions were inflexible.

Many tour operators adapted their business to selling international tour product well in advance of Covid-19 and many have pivoted their product offerings to domestic or short-haul international travel. Even with the advent of travel corridors/bubbles applying to some regions of the world, the marketing options available to many tour operators have been severely restricted.

Travel Agents and Online Booking sites

For many travel agents (excluding those who focus on business clientele), risk management tended to focus on the naïve assumption that if you can't sell one destination you simply focus your attention on another. Of all sectors in the travel industry, travel agents and their younger cousins, the online booking agencies, were the least prepared sector in the travel industry to deal with a global crisis on the scale of Covid-19.

Within a short time, in many countries, travel agents (especially the majority who relied on selling international travel) found themselves with little or no product to sell. Although they could sell domestic travel in many countries, the majority of travel agents lacked the skill, knowledge and motivation to do so. A minority of travel agents have refocused to domestic travel with considerable success. For many travel agents, much of their time was devoted to processing refunds and several were exposed by the media and government authorities for charging excessive 'refund processing fees' on top of the refunds provided by principals.

High street bricks and mortar travel agencies (travel agencies with a physical presence in shopping malls or retail centres) rapidly found their businesses to be unviable in an environment in which they had little to sell and most transactions involved refunds. Online booking agencies, home based or mobile travel agents, unencumbered by business property rental fees for space in shopping malls and retail precincts, were able to survive longer without rental overheads. However, travel agencies (most of which are small businesses) were becoming increasingly dependent on government support to retain their staff.

Travel agency conglomerates, heavily reliant on 'marketing support' from 'preferred principals' suddenly found that sources of income evaporated, when preferred principals themselves were starved of cash flow.

Many travel agencies, unable to adapt to the massive business changes in their marketing environment wrought by Covid-19, struggled to maintain their existence. Many outlets closed by the conglomerates during the pandemic are unlikely to re-open unless they radically alter their business strategies, which involves marketing and selling viable product and services or operating in hibernation while preparing for a resumption of international business.

DESTINATION MARKETING AND MANAGEMENT ORGANISATIONS

At a global level, the UNWTO, WTTC and the Pacific Asia Travel Association (PATA) and the Asia Pacific Economic Cooperation (APEC) Tourism Working Group, and regional associations such as EU Tourism, Caribbean Tourism Association and the Association of Southeast Asian Nations (ASEAN) Tourism have devoted considerable attention to risk, crisis and recovery management for some years. This was especially noticeable after the 9/11 attack in September 2001. One of the most effective tourism recovery responses to pandemics was developed by PATA under the label of Project Phoenix in 2003 in response to SARS (Yates, 2006). In essence, this campaign involved an alliance between media, government owned tourism marketing organisations and the private sector of the Asia Pacific tourism industry to restore the reputation and positive safety perceptions of Asia Pacific tourism destination. This is discussed in more detail in Chapter 7 of this book. Although Project Phoenix proved to be an effective tourism recovery template in response to SARS and many subsequent pandemic events, its efficacy was limited during Covid-19 largely due to the imposition of government regulations that restricted international travel. Despite the increased attention paid by destination marketing and management organisations (DMOs) to respond to pandemics they were largely unprepared to the extreme scale of global challenges imposed on tourism by Covid-19.

In 2005, the Japan National Tourism Organisation launched a new marketing campaign called Yokoso. The campaign assumed relevance far greater than its authors could have ever imagined. The Yokoso marketing campaign, then, became the flagship strategic marketing campaign of Japan for the next 15 years (Uzama, 2012). It was designed to achieve three key strategic goals.

1. Increase the number of international arrivals into Japan.
2. Increase domestic tourism within Japan.
3. Reduce the number of Japanese travelling internationally.

In 2020, these issues struck a chord with many destination tourism marketers seeking a strategic approach in response to the Covid-19 pandemic.

Covid-19 has posed even greater challenges to the world's destination marketing bodies, most of which are funded by national, state/provincial/prefectural governments or local governments. As government funded bodies, DMO's are relatively free of commercial concerns, which limit the private sector's scope of action, but Covid-19 has placed such heavy pressures on government budgets that tourism DMOs struggled to attract the funding they needed to operate effectively in a challenging marketing environment.

Most national tourism organisations tended to focus the bulk of their marketing activities on attracting international visitation and prioritise source markets based on visitation numbers, length of stay, per capita expenditure and the propensity of travellers from that source market for repeat and referral visitation. The majority of local tourism DMOs rely on domestic visitation and state/provincial/prefectural DMOs seek a mix between this and international visitation.

The widespread closure of international borders, in response to Covid-19, forced most national DMOs to reassess their marketing and promotional strategies to focus on domestic tourism. Some countries in which governments had conditionally opened borders to selected nationalities could maintain international marketing activities to selected source markets. However, the reduction of international transport links, along with consumer concerns about the safety of travel, meant that it was not always possible to rely on a rapid resurgence of international tourism.

All DMOs were affected by the proliferation of negative government travel advisories applied to a wide range of destinations in response to the pandemic. The governments of many countries including China, Thailand, Australia and New Zealand banned international travel for their citizens apart from a small number of exempt categories. In China, Thailand and Australia, government directives represented a ban on international travel but for most countries' government travel advisories, not to travel to destinations were framed as advice. Many more countries have either advised their citizens to avoid certain destinations or issued cautionary advisories applying to those countries perceived as high risk for Covid-19. Negative or cautionary advisories are usually reflected by the decisions that travel insurers make when offering travel insurance coverage to certain destination.

Case Study 3.1

Tourism Australia's Strategic Response to Covid-19

Tourism Australia's (TA) strategic response to Covid-19 has been widely regarded as a global best practice model for the response of national DMOs to the pandemic. Australia's recovery strategy represents a strategic response to Covid-19 and is relevant to many DMOs worldwide.

Tourism Australia, in common with most DMOs, responded to negative perceptions of an entire destination which arose from localised crisis events. In Australia, the final months of 2019 and the beginning of 2020 was dominated by massive and highly destructive bushfires, which caused significant property, environmental damage and loss of life in many parts of South-East Australia. The bushfire outbreak was Australia's most extensive bushfire event in over 200 years. The states of New South Wales (NSW), Victoria, South Australia, South-East Queensland and the Australian Capital Territory (ACT) (where the national capital, Canberra is located) were the most severely affected areas.

TA's challenge was to communicate the distinction between the international perception that much or all of Australia was burnt and the fact that while parts of Australia had indeed sustained fire damage, most tourism attractions were open and operating.

(Continued)

It is notable that the extent of this impact was felt far beyond the regions directly and physically affected by the fires. Most major Australian cities (including Sydney and Melbourne) were spared direct fire damage but experienced smoky skies during much of December 2019 and January 2020. Many of the more apocalyptic perceptions were fed by international media coverage of the Australia's bushfires. TA's own research of international markets revealed that over 60 per cent of consumers believed that more than 25 per cent of Australia was burnt (TA, 2020).

Initially, TA was slow to respond to the bushfire challenge and embarrassing PR errors occurred. The most notable marketing *faux pas* was the screening of a happy go lucky promotional video (*Matesong*). The video invited British citizens to escape the travails of Brexit and the British winter to visit sunny, happy go lucky and matey Australia. It was first screened for the UK market on 25 December 2019, when the Australian bushfires were at their height and many of the sites depicted in the video had been destroyed by the fires. The inappropriate timing of its release (more pertinent than the quality of the content) was an expensive PR debacle for TA. Recent changes in TA's management prior to the fires meant no staff member was given responsibility for risk and crisis management. For its part, TA (2020) maintained that the peak of the fire occurred a week after the screening of *Matesong* at which time the advertisement was withdrawn. This changed in January 2020, with the appointment of a staff member assigned responsibility for risk and crisis management. The appointee, Elisa Jakymin, was a former student majoring in tourism at the University of Technology Sydney (UTS). Her crisis management skills were put to the test during the Covid-19 outbreak.

Elisa, in association with Rob Dougan, General Manager, Strategy and Research, was involved with a number of TA colleagues, including CEO Philippa Harrison, Corporate Affairs Manager, Bede Fennell, and Chief Marketing Officer Susan Coghill, in developing an effective strategy to contextualise the factual extent of the fires. This team was instrumental in developing TA's recovery strategy in response to Covid-19. In doing so, this would counter the perceptual impact based on international media coverage, which depicted Australia as a country immolated by the fire. TA also used the fire damage as a call to Australians to support fire ravaged communities (provided those communities agreed) by visiting them – an effective employment of solidarity tourism.

Tourism Australia's 'Holiday here this year' campaign was launched on 23 January 2020, a few days after the Australian government announced an AU$76 million 'rebuilding Australia tourism' package to restore tourism. Some of this funding was directed to international markets and about half was used to promote domestic tourism. Although the funding was initially in response to the bushfires, it was used to develop marketing campaigns in response to Covid-19 (TA, 2020).

As the bushfire crisis subsided by late January 2020, the Covid-19 pandemic followed almost immediately. International inbound and outbound tourism was the first casualty. Between January and March 2020, the Australian government rapidly banned inbound tourism from a growing number of countries starting with China but extending travel bans for citizens from all countries apart from returning Australian citizens, permanent residents and their dependents, and those with special permission. Conversely, the Australian Department of Foreign Affairs issued travel advisories, advising Australian not to travel to any country in the world. During April 2020 international visitation to Australia was 99.7 per cent down on April 2019 (Australian Bureau of Statistics, 2021). This made any active TA international marketing untenable, in the short term.

A complicating factor for TA's response to Covid-19 centred on the significant differences between Australia's six states and two territories in terms of interstate border controls and the social distancing regulations, applicable to each state. In addition to Australia being effectively closed to international tourism, interstate domestic tourism was disrupted. In addition, there were significant barriers within Australia (including blockades and entry quarantine requirements) restricting travel from one Australian state to another. Tourism Australia sought to consolidate messaging and announcements regarding state borders via the creation of a travel status map on its corporate website (TA, 2020). The lack of uniformity in tourism access and regulations within Australia severely hampered any meaningful negotiation of international travel bubbles/corridors. Although some tourism industry associations (Australian Chamber of Commerce and Tourism, 2020) had proposed Australia–New Zealand travel corridors confined to specific Australian states or territories, the administration of such travel corridors would be difficult. Australian passport holders are not required to state their place of residence in their travel documents.

TA's Covid-19 strategy involved the following core elements:

1. Focus on the promotion of local and domestic travel pitched to the Australian market. This was an area of tourism promotion little emphasised prior to the 2019–2020 bushfires and Covid-19. While a growth in domestic tourism was not expected to replace the loss of international tourism, it was hoped that it would partially fill the financial vacuum.
2. Due to the closure of some interstate borders within Australia, focus on inviting Australians to get to know their local area and regions within their state/territory.
3. Solidarity tourism, with an emphasis on supporting regional destinations affected by the bushfires. TA allocated AU$61 million of the bushfire funding package to support bushfire-affected regions split between international and domestic markets.
4. Promote healthy travel practices. Although this was not a core activity of TA, its weekly webinars provided a platform for government medical officials to brief tourism industry professionals on health-related matters about Covid-19.
5. Understanding programmes and access to government financial support available to tourism professionals and enterprises during the pandemic. TA provided this information on its corporate website and it featured in a number of its webinars. There was close collaboration between TA and the various state tourism organisations to release information in each Australian state, which would highlight the support available from each state or territory government.

(Continued)

6. Coping with psychological strategies in response to the three key stages of the crisis: panic, learning and resilience.
7. Promote and advocate the idea of viable international travel bubbles (initial emphasis focused on New Zealand and the south-west Pacific).
8. Promote and advocate priority cohorts of tourism for early re-entry. These included young travellers on working holiday visas who provided a significant portion of seasonal agricultural labour and who visited many regional areas during a lengthy stay. Universities and tertiary institutions similarly advocated priority for international students.
9. Maintain international interest in Australia as an aspirational destination to visit when the pandemic ebbs.
10. Be responsive to changes in the regulatory environment and maintain governmental awareness of tourism industry concerns.
11. At all times, talk up long-term resilience of tourism generally and Australia specifically, and in doing so, maintain stakeholder communication.
12. Provide resources and opportunities to train Australian and international travel professionals to effectively identify, market and sell Australian tourism products and experiences. TA's Aussie Specialist Programme, which has been designed to train international travel consultants to understand and sell Australian travel products, was adapted for Australian travel professionals.
13. Maintain the interest and knowledge of international stakeholders about Australian tourism in order to be prepared for an eventual resumption of international tourism.
14. Work cooperatively with all sectors of the tourism industry to pivot to changes in the marketing environment that would facilitate a resumption of widened domestic and international tourism.

From March 2020, TA ran an extensive series of weekly national and global webinars, primarily targeting travel industry professionals, which provided a platform for its management, led by CEO Philippa Harrison, travel industry association leaders, key figures from all sectors of the Australian travel industry, including the CEO of Qantas and the CEO of Sydney Airport. As a national government entity, TA's webinars featured Prime Minister Scott Morrison, Tourism Minister Simon Birmingham, treasury officials (to discuss financial support measures), national health officials (who could explain Covid-19), and a range of independent management and marketing experts.

Tourism Australia benefited from a significant increase in government funding for Australian destination marketing, approved early in 2020 towards the end of the bushfire crisis. A second factor was that the 11 million Australians who travelled internationally in 2019 had no alternative means of fulfilling their desire to travel throughout most of 2020 except via domestic travel.

Australia, in common with Japan in 2005, had a negative trade balance in tourism. In 2019, 9.4 million international tourists visited Australia, spending AU$44 billion dollars, while 11.2 million Australian travelled internationally, spending AU$68 billion. A benefit arising from the Covid-19 pandemic would mean that the negative tourism balance was reversible.

A staged recovery is recommended AS PART OF AN INDUSTRY RESPONSE

	CRISIS		RECOVERY	NEW NORMALITY
	1. Panic (SURVIVAL & PROTECTION)	2. Restricted movement (PREPARATION)	3. Rising optimism (SHORT TERM RECOVERY)	4. Free movement (LONG TERM RECOVERY)
DESCRIPTION	Fear is stopping us from making any plans	Forced social isolation / closed borders and communities	Relaxation of social isolation, economy restarting, travel intent rising	Travel kicks off again and self isolation halted
CRITERIA	Border closures Negligible travel interest	Border closures research travel interest Negligible booking interest	Border closures research travel interest Booking interest Negligible intent to travel	Local / Aussie borders open Air capacity available Interest to book and travel Forward bookings
MINDSET	Fear and trepidation	Frustration and boredom	Exasperation with optimism	Enthusiastic and adventurous

Figure 3.4 Tourism Australia's strategic approach to recovery during the Covid-19 crisis

Source: Elisa Jakymin, Tourism Australia. Used with kind permission of Tourism Australia.

Figure 3.5 Tourism Australia's 2020 slogan for the Australian market, a message that applied to many countries affected by the Covid-19 restrictions on international travel

Source: Tourism Australia, 2020. Used with kind permission of Tourism Australia.

(Continued)

The tourism sector has been severely impacted by Covid-19 restrictions

➤ **Tourism is an important and growing sector of Australia's economy**

- ✓ **Expenditure**: $152 billion in expenditure
- ✓ **GDP**: 3.1% of national GDP
- ✓ **Employment**: 5.2% of total jobs (666,000)
- ✓ **Businesses**: 300,000 businesses; 100,000 in regions

➤ **Covid-19 has hit the tourism industry hard**

- **Employment** – **1 in 3 jobs lost** in tourism
- **Expenditure** – **down 40%** for 2020–21
- **Businesses** – **60%** of accommodation businesse on JobKeeper
- Intl spend – **negligible** for June & Sep 2020 quarters ($8.2 billion in June 2019; $11.1 billion in Sep 2019)
- **Global travel** – **down 60–80%** in 2020, Australia border restrictions likely until mid 2021
- Demand in tourism industries **down 95%** (BCG)

➤ **There are many tourism-reliant regions at risk**

- **44 cents** in every tourism dollar spent in regions
- **47% of tourism jobs** are in regional Australia
- In 8 regions, **25–50% of jobs** are in tourism
- 15 of Australia's tourism regions have suffered **> 30% drop in tourism expenditure**

Figure 3.6 Tourism Australia's logical progression for its Covid-19 strategy

Source: Tourism Australia, 2020. Used with kind permission of Tourism Australia.

The resources and capabilities to support TA's agenda benefited from a supportive Prime Minister. Prime Minister Scott Morrison was a former Managing Director of TA in 2006–2007 and had extensive professional involvement in tourism prior to his political career. More than most former Australian prime ministers, he understood the significance of tourism from the economic, national image and employment creation dimensions.

Government economic support for employees and business in all sectors of the Australian economy affected by Covid-19 enabled enterprises to draw on extensive government funding support through the 'Job Keeper' programme, which involved payments to companies designed to be passed on to employees who would otherwise have been dismissed. The Job Seeker programme supported unemployed people attempting to return to the workforce. In both cases, payments to recipients was far more generous than previous social services payments. At a time when many tourism and hospitality businesses experienced a massive loss of revenue, the subsidies retained many jobs and have bought time for many tourism businesses to make adaptations to their business strategies. Job Keeper, initially guaranteed until September 2020, was extended at a reduced level until the end of March 2021.

These measures, valuable as they were, couldn't fully prevent many tourism and hospitality businesses in Australia from closing their doors. However, TA (backed by Australia State Tourism Boards) effectively used its position as a government agency and its alliances with all key sectors of the Australian tourism industry to advance a collaborative approach to managing and recovering from the Covid-19 crisis.

A GLOBAL APPROACH TO COVID-19 TOURISM RECOVERY

Many of the strategies adopted by TA in response to Covid-19 were similar to those used in other parts of the world. Certainly, the focus on local and domestic travel and the building of tourism industry resilience is widely employed. The governments of many countries have provided support to maintain the viability of tourism enterprises, operations and employees. Some countries, notably most EU countries, and many Caribbean countries opened their borders to limited international tourism by mid-2020. Many countries are operating travel corridors/bubbles, showing they are well ahead of Australia. The PATA Covid-19 Tourism Recovery Plan (Figure 3.7) bears many strategic similarities to TA's strategy. PATA's approach is also similar to the strategic direction advocated by the UNWTO. In late May 2020, the WTTC released its *Safe Travel Protocols*, which was a bullet point guide for Covid-19 safe practices with a specific focus on each sector of the tourism industry (WTTC, 2020).

TOURISM AND HOSPITALITY TRAINING AND EDUCATION

It would be remiss of the author as a tourism educator to ignore the important impact Covid-19 has had on tourism education and training. During 2020, the debates about whether teaching of tourism should be conducted online effectively became moot. Virtually all educational and training institutions adapted to online teaching and learning once it was realised that social distance regulations, localised lockdowns and travel restrictions made the safe delivery of live teaching difficult, if not impossible.

Fortunately, the online teaching technology used by many institutions for years were adopted by virtually all educational and training institutions. For many education and training providers, face-to-face or live training and teaching is a preferred option (especially in the hospitality sector) but as the experience of thousands of

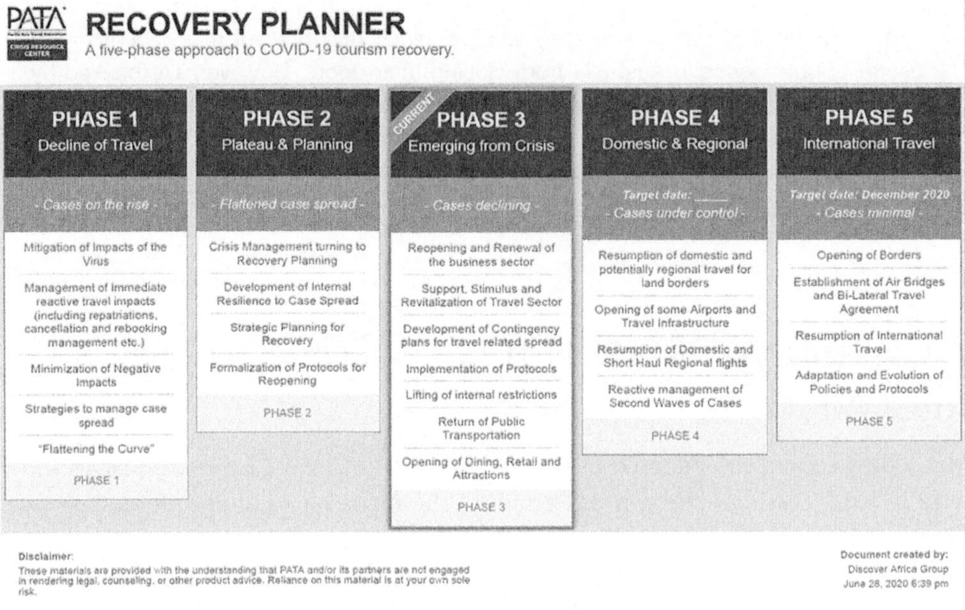

Figure 3.7 PATA's strategic approach to Covid-19 tourism recovery

Source: Pacific Asia Travel Association with kind permission of PATA, 2020

educators and students showed in 2020, online teaching and training is viable. In fact, it had some advantages for teachers and students alike. Lecturers and teachers could source guest speakers from anywhere in the world provided they were accessible online. Students lost some of their best excuses for non-attendance in classes as they could literally participate in classes from the comfort of their bed. Conversely, student teamwork was complicated somewhat when students only knew each other from their small images on Zoom, GoToMeeting or other meeting and classroom site.

The other major change to the education and training environment has been the massive changes to the content of tourism education wrought by Covid-19, discussed at length in this and many other new books and journal publications. Certainly, as most leading tourism journal editors have clearly stated, tourism research from 2020 that ignores the impact of Covid-19 has limited value.

Until the Covid-19 pandemic is contained through a vaccine or treatment regime, there has been a massive upsurge in training, intellectual exchanges, conferences and teaching units developed and presented online. This has reduced the cost of a great deal of excellent educational and training products and enabled large numbers of participants to share and receive knowledge from global authorities in tourism.

A significant question, yet to be answered, is whether and to what extent Covid-19 and the ravages it inflicted on all sectors of the tourism industry will influence future demand for tourism education and training.

The prevailing school of thought is that tourism will recover, although there is extensive debate on the duration and extent of that recovery. Certainly, in the short term, there has been a massive loss of jobs and in the space of a few months in 2020, tourism shifted from an industry with a skills shortage to one of a job shortage. The choice to enrol in tourism education and training is linked to available employment opportunities for graduates of educational and training programmes. It is unrealistic not to expect a short-term drop in demand for tourism training and education enrolments, and its associated impacts on employment in education and training programmes and available teaching and training positions. In turn, this may affect the viability of some training and educational institutions and providers. On the plus side, there is a likelihood that those positions that remain available will be more amenable to choosing applicants with reputable tourism and hospitality qualifications. This may make students adopt a more considered approach to choosing education and training institutions, which focus on quality rather than those merely issuing degrees and diplomas in exchange for cash.

THE ALTERED PARADIGM OF GLOBAL TOURISM AS A CONSEQUENCE OF COVID-19

The tourism industry has been forced to adjust to a new set of realities in response to Covid-19. The pandemic has clearly demonstrated that the free movement of people around the world is no longer a guaranteed right but dependent on safety in its broadest sense. This of course includes health and safety. Tourism's economic vulnerabilities were exposed and while tourism is likely to eventually recover as an industry, those countries traditionally relying on tourism as the mainstay of their economies are now aware they need to diversify their economies and find alternative avenues to employ their people. While tourism will return for the most part to be a positive force in the world, the Covid-19 outbreak has given many tourism planners, policy makers and the industry an opportunity to reflect on some of the negative practices of pre-pandemic tourism.

For some destinations such as Venice, Majorca and Barcelona, the instant transition from over-tourism to zero tourism was a case of overkill. Sustainable tourism has long been a major consideration in global tourism but talk needs to translate into action. The achievement of a sustainable balance in carrying capacity would be a very positive post-pandemic legacy.

Much of the global environment has literally taken a breather due to the pandemic. Pollution abated in many of the world's cities. The precipitous reduction in air, sea and land traffic reduced human pollution of the skies, sea and the natural environment.

The pandemic effectively confounded the paradigm of endless tourism growth. Even the most optimistic forecasters accept that tourism is likely to take three years to return to 2019 levels and most tend to think it will take five years. While the aspiration to resume travelling will understandably remain strong, the trauma of Covid-19 and the possibility of a similar event occurring in the future is likely to be a restraining factor in converting aspirations into reservations for some time.

A POST-PANDEMIC TOURISM BUSINESS AND RECOVERY

For most of the past 50 years, tourism businesses have operated on an illogical paradigm that volume of business is more important than yield. Competitive pressures to keep prices low and maximise capacity of airline seats, hotel rooms and cruise cabins have overridden business logic throughout the tourism industry. Airlines, tour operators, cruise operators, travel agents, accommodation providers, event organisers and attractions have generally adopted the business model based on high turnover–low yield perishable tourism. This usually works when turnover is high, but low profit margins result in low reserves when a business experiences a downturn. The majority of tourism enterprises lack the reserves to ride out a business slump on their own resources. In many countries affected by Covid-19, governments helped them and their employees out with financial support (in many other countries the businesses collapsed). Tourism business need to operate on an altered business paradigm, remain competitive but operate with higher yield and ensure that there is sufficient profit reserve to maintain the business for downturns.

Governments, even those with the best intentions, should not feel obliged to bail out irresponsible business owners and companies unless they are vital to a nation's strategic interests. Running a business on a sustainable economic basis does not necessarily translate into high and unaffordable prices for consumers, but it is more than likely that tourism businesses making the choice to operate as viable businesses may have to increase prices for their travel products and services.

A more commercially responsible business management on the part of business enterprises may slow the pace of tourism's democratisation into the future but it does not mean the process will stop. Tourism businesses need to emphasise the value proposition of their services ahead of a low price. Safety (including health safety) will almost certainly become a more prominent determinant of product and destination choice, especially in the immediate post-pandemic environment. Safety and health measures cost money and should be factored into pricing.

The key beneficiary of resurgent tourism demand, short term, will be local and domestic tourism. This has been a common response to many tourism crises.

Restoring international tourism is almost certain to be gradual. Travel bubbles/ corridors and regional travel zones will be determined by governments, rather than consumer perceptions and demand. Carrying capacity and spacing applying to all forms of tourism activity, including events, cruises, coaches, trains, airlines, attractions, conventions, accommodation, restaurants etc., are likely to be determined by governments.

The cruise sector faced intense scrutiny from governments and the media in the wake of Covid-19. Cruising will need to balance onboard hedonism with responsible health practice and while many of the cruise sector's core market will return, cruise operators are obliged to visibly address health and safety concerns to family travellers and younger travellers who were recent adherents of cruising.

Two sectors of tourism likely to be altered by Covid-19 are business travel and educational travel. There is no doubt that both will continue but in a reduced form. During the pandemic, businesses learned that many of their transactions and meetings were effectively and inexpensively conducted online. While some cultures insist on a face-to-face meeting to open or close a deal, or inspect operations with a company, many businesses are likely to treat future business travel as a perk rather than a necessity. A major challenge for business travel managers will be to convince many businesses who have shared the financial travails of the tourism industry inflicted by Covid-19 of the relevance and value of business-related trips. Many of the same issues apply to the education sector. Educational institutions, which experienced significant financial strain due to the pandemic, are almost certain to impose strict justifications on educational and research-related travel. The attraction of a tourism career for prospective students and trainees in educational and training institutions was compromised by the Covid-19 experience. These institutions need to provide a compelling case that the long-term future of tourism can guarantee a career after graduation.

Tour operators face the likelihood that large tour groups will either be split into smaller sub-groups or replaced by small tour groups. Carrying capacity at attractions, hotels and dining venues and social distancing will remain a major influence on the operation of tours well after Covid-19.

The travel industry will need to re-evaluate its approach to risk management. Traditionally, the approach to risk and crisis management in tourism was based on the expectation of a snap-back recovery from a crisis. Covid-19 is revealing that a V-shaped recovery is most unlikely. The key issue for all tourism business is to ensure that they are prepared for a long-term crisis. The lack of preparedness for Covid-19 was evident across all sectors of tourism. Travel agents (especially traditional bricks and mortar travel agents) were exposed as especially vulnerable businesses. Many of them were too smug to treat business survival seriously and too slow to adapt to rapidly changing challenges imposed on them by Covid-19.

One sector that has been significantly compromised by Covid-19 is the MICE market. Meetings, conventions and exhibitions need to adapt their planning, management and operations to meet with heighted expectations of a healthy and safe environment. In both the MICE sector and the accommodation sector, managing events and food and beverage services will need to adhere to stringent health and safety regulations. The stand-up or sit-down buffet may be consigned to a happy memory for consumers and businesses.

Tourism needs to make a case for its long-term viability as a reliable contributor to national economies in both good and adverse time. The pandemic crisis exposed the real fragility of the tourism industry. The post-pandemic challenge for the tourism industry is to re-image itself as resilient, economically sustainable and socially and environmentally sustainable. The great hope for tourism comes from the Mandarin root of 'crisis' – danger and opportunity.

There is a real opportunity for post-pandemic tourism to be more sustainable. Over-tourism has the potential be replaced by manageable tourism. Cheap tourism may be superseded by high quality, value-for-money tourism. Covid-19 has forced tourism businesses to operate in an economically responsible and sustainable manner. Tourists can be encouraged to enjoy and experience a growing range of less-crowded destinations. Covid-19 has taught travellers and tourism professionals alike to treat tourism as a privilege, not as a right. Ironically, the most severe crisis in modern tourism history may result in the treatment of tourism as an experience to be treasured and respected, rather than abused in a haze of inebriation, selfish hedonism and irresponsibility. While most tourists are responsible and respectful guests in the destinations they visit, the downside of travel being available to all means that tourism has attracted the best and worst of tourism generating societies.

Discussion Questions and Activities

1. The author has outlined a number of the elements that make the Covid-19 crisis unique. What do you see as some other elements that are unique, which have not been emphasised in the chapter?
2. The Tourism Australia case study focused on a staged post-Covid-19 recovery – from local, domestic to staged international travel recovery. Focusing on a tourism generating region of your choice, how would you envisage a post-pandemic tourism recovery would play out?
3. Choose a tourism industry sector and discuss a strategic approach to a post-Covid-19 recovery.

4. Choose an individual tourism/hospitality/event enterprise or destination and develop a post-Covid-19 recovery strategy.
5. What do you believe are the key lessons the tourism industry has learned from Covid-19 and how would these lessons be applied to a destination, industry sector or an individual tourism/hospitality/event enterprise of your choice?
6. The author has outlined a range of paradigm shifts for global tourism in response to the Covid-19 pandemic. Either individually or as a group, develop your own set of pre-pandemic and post-pandemic paradigms for global tourism.

A BRIEF POSTSCRIPT, MAY 2021

By the end of April 2021, the official number of global Covid-19 cases exceeded 150 million and the number of deaths surpassed 3.2 million. Despite the fact that extensive vaccination campaigns employing several different vaccines had commenced in many countries, the daily number of new cases remained above 400,000 (80,000 in the US alone). From April 2021, the number of cases increased exponentially in India and according to many observers the real rate of infection and death in India is likely to be 10 times that of official government statistics. The nature of the pandemic had become more complex as new and more virulent strains of Covid-19 appeared in the UK, Brazil, India and South Africa. Although governments around the world have high expectations that mass vaccinations will ultimately contain the spread of the pandemic, it is clear that there is no quick fix and the global tourism industry is preparing for a slow recovery, which may involve several years. Many governments all over the world have maintained a reluctance to re-open borders for international tourism free of quarantine restrictions and it is likely that pre-arrival testing and evidence that tourists have been vaccinated are likely to be core conditions attached to international travel. The WHO's mission to Wuhan, China, which took place during January and February 2021 to establish the origins of the pandemic shed little light on the origins of Covid-19. Irrespective of the true origins of Covid-19 there is no doubt that the pandemic and associated government restrictions have imposed a tyrannical grip on global tourism which will take years to overcome.

REFERENCES

Airports Council International (2020) https://aci.aero. Accessed 19 December 2020.

Australian Bureau of Statistics (2021) Tourism Arrivals to Australia Update February 2021. https://www.abs.gov.au/statistics/industry/tourism-and-transport/overseas-arrivals-and-departures-australia/latest-release. Accessed 7 May 2021.

Australian Chamber of Commerce and Tourism(2020) *The Trans Tasman Bubble: A Beacon for Tourism and Trade*. https://www.australianchamber.com.au/trans-tasman-bubble-a-beacon-for-offshore-tourism-and-trade. Accessed 7 May 2021.

Avraham, E. and Ketter, E. (2016) *Tourism Marketing for Developing Countries Battling Stereotypes and Crises in Asia, Africa and the Middle East*. Palgrave Macmillan: Basingstoke, UK.

Beirman, D. (2016) (Updated in 2019) *Tourism Risk, Crisis and Recovery Management Guide*. Council of Australian Tour Operators: Sydney, Australia.

Beirman, D. (2019) 'Collaborative approaches to government travel advisories in Australia between Australia's tourism industry leadership and the Australian Department of Foreign Affairs and Trade', *Journal of Vacation Marketing*, 25(1): 71–87.

Beirman, D. (2020) 'The trouble with travel bubbles', *New Europe*, 29 July 2020.

Braun, S. (2020) *Latest Conspiracy Theories Repeat Old Stories*, Deutche Welle, 29 May 2020. https://dw.com/en/latest-conspiracy-theories-repeat-old.storieds/a-53603358. Accessed 3 December 2020.

Centers for Disease Control and Prevention (CDC) US Government (2020) www.cdc.gov. Accessed 20 December, 2020.

Cruise Lines International Association (CLIA) (2020) https://cruising.org/en/. Accessed 2 December 2020.

Cruise Lines International Association (Australia) (2020) www.cruising.org.au. Accessed 2 December 2020.

Elsegood, S. (2019) *The Airline Industry 2019*. (Simon Elsegood is the Global Research Manager Centre for Aviation (CAPA).) Presentation to UTS students, studying Managing Tourism Sectors. 9 August 2019 (unpublished).

Gossling, S., Hall, C.M. and Scott, D. (2020) 'Pandemics, tourism and global change. A rapid assessment of Covid-19', *Journal of Sustainable Tourism*, 29(1): 1–20.

Higgins-Desbiolles, F. (2020) 'The war over tourism. Challenges to sustainable tourism in the tourism academy after Covid-19', *Journal of Sustainable Tourism*. Online. https://doi.org/10.1080/09669582.2020.1803334. Accessed 6 August 2020.

International Air Transport Association (IATA) (2020) www.iata.org. Accessed 2 December 2020.

New South Wales Government (2020) *Special Commission of Enquiry into the Ruby Princess*, 14 August 2020. Government of NSW: Sydney, Australia.

Nirenstein, F. (2020) 'How the coronavirus sparked anti-Semitic conspiracy theories', *Jerusalem Centre for Public Affairs*, 21 April 2020. Jerusalem, Israel.

Oriel, J. (2020) 'Pandemic or pandemic. The conspiracy theories abound', *The Australian*, 14 August 2020.

Pacific Asia Travel Association (PATA) (2020) www.pata.org. Accessed 3 December 2020.

Royal Caribbean Cruise Lines (2020) https://www.royalcaribbean.com. Accessed 28 November 2000.

Secon, H. and Woodward, D.A. (2020) 'Comprehensive timeline of the new coronavirus pandemic from China's first case to the present', *Business Insider*, 20 March 2020.

Thiessen, T. (2020) 'Europe travel. Good and bad news about where you can visit', *Forbes Magazine*, 15 May 2020.

Tourism Australia (2020) www.tourism.australia.com. Accessed 02 December 2020.

United Nations World Tourism Organisation (UNWTO) (2020) www.unwto.org. Accessed 2 December 2020.

Uzama, A. (2012) 'Marketing Japan's travel and tourism industry to international tourists', *International Journal of Contemporary Hospitality Management*, 21(3): 356–65.

Vargas, N. and Equinas, S.A. (eds) (2020) *Covid-19 Airway Management and Ventilation Strategy for Critically Ill Older Patients*. Springer: New York, USA.

World Health Assembly (2020) www.who.int/about/governance/world-health-assembly. Accessed 3 December 2020.

World Health Organisation (WHO) (2020) http://www.who.int. Accessed 2 December 2020.

World Health Organisation (WHO) (2021) Interactive Timeline. https://www.who.int/emergencies/diseases/novel-coronavirus-2019/interactive-timeline. Accessed 7 May 2021.

Worldometers (2021) *Statistics on Coronavirus*. https://www.worldometers.info/coronavirus. Accessed 3 May 2021.

World Travel and Tourism Council (WTTC) (2020) *Protocols in the New Normal*. www.wttc.org. Accessed 30 November 2020.

Yates, M. (2006) 'Project Phoenix: A benchmark for reputation management in travel and tourism', in Wilks, J., Pendergast, D. and Leggat, P. (eds), *Tourism in Turbulent Times: Towards Safe Experiences for Visitors*. Elsevier: Amsterdam. pp. 263–76.

PART 2

Key Themes in Tourism Risk, Crisis and Recovery

4

THEME 1: POLITICAL INSTABILITY AND TOURISM

INTRODUCTION

There has been a plethora of both academic and travel industry research that indicates the primary factor in destination choice for tourists is safety. This book addresses specific threats to safety arising from terrorism, crime, natural disasters and epidemics/pandemics. However, at the root of many threats to tourism safety and security is the absence of reliable institutions of legitimate constitutional government and governance, law and order, emergency management, healthcare and legal protection.

For the purpose of this chapter, political stability involves the ability of a national government to access or compete for power through peaceful political processes. This involves the people under that jurisdiction to contribute to benefit from services and security provided by the state. The eight pillars of stable government are (Gerschewski, 2013; Wahab, 2019):

1. The ability to provide essential services, including defence of the state and internal law, order and security.
2. Stewardship of state resources for the benefit of all citizens.
3. Civic participation and empowerment.
4. Government officials are accountable through political and legal processes.
5. A consistent range of civil rights and legal obligations, which include: the provision of basic education, protection of minority rights and a guarantee of the protection of basic material and human needs for all citizens.
6. The ability to control its borders from external threats.
7. The operational integrity of the banking system and currency exchange rates.
8. Certainty of policy.

While most of the above tend to be associated with parliamentary democracies, all governments (including totalitarian regimes) are fundamentally accountable to their citizens. The legitimacy of any government depends on acceptance, whether by popular assent or by the de facto acceptance by the majority of citizens of its rule. Historically, even the most coercive and militarised dictatorships can and have been overthrown by popular revolt or a military *coup d'état* if they lose legitimacy or their ability to exercise coercive authority over their citizens. In recent times, the collapse of the Soviet Union and communist regimes in Eastern Europe in the late 1980s and early 1990s are stark evidence that one-party coercive rule could not survive indefinitely without, at minimum, the tacit acceptance of the people.

THE TOURISM PERSPECTIVE OF POLITICAL STABILITY

From the perspective of visitors to a country, an assurance of law and order, reliable institutions to enforce safety and security, access to professional healthcare and peaceful social, economic and political environments are minimum requirements to ensure that a tourist can visit a country in safety and with confidence. Some countries, due to a range of long internal conflicts, have degenerated into failed states. Somalia, South Sudan, Afghanistan and Venezuela are a few of the countries in which internal disorder is so chronic that the country is virtually ungovernable, and some of that internal anarchy creates an external threat. In the case of Somalia, over 20 navies patrol the Gulf of Aden to prevent sea piracy. For Somalian coastal communities, piracy emerged as the only income source they have.

For most of the world's 1.4 billion international tourists (UNWTO, 2018), the concern for safety and security does not involve a proscriptive preference for travelling to countries governed in accordance with a particular political system. Certainly, for a minority of travellers, the decision to visit a country may be influenced by approval or disapproval of that country's form of government or its political regime. A number of political and religious activists, academics and journalists have advocated boycotts to specific countries as a means of expressing disapproval for a specific regime and its policies. At times, the attempt to impose a travel boycott or restrict travel to a specific country may occur at government level. Many Arab and some non-Arab Muslim countries routinely denied entry to Israeli passport holders or travellers with evidence in their passports of a visit to Israel. The 2020 *Abraham Accord* peace agreements between Israel with Bahrain, the United Arab Emirates, Morocco and Sudan demonstrate these boycotts are far from permanent. Generally, a government-imposed travel ban or boycott to another country applies to travel between countries that have no bilateral diplomatic relations or when bilateral relations are defined by a state of mutual belligerency or mutual non-recognition. The Covid-19 outbreak of 2020 was exceptional because many countries wholly or selectively banned outbound international travel on the grounds of health concerns.

During South Africa's Apartheid regime (1948–1994), many countries, especially those in the former Soviet bloc and many African governments, denied entry to South African citizens and travellers from other countries with a South Africa visa. From 1959 to 2015, the USA imposed a travel boycott against Cuba's communist regime. These issues are not matters related to political stability but represent political barriers to tourist visitation. In today's global environment, these restrictions had relatively minor impacts. Visa restrictions against Israel were well known and legally circumvented by many travellers.

However, most travellers' destination preferences are guided by the perceived stability of the country's political system. A tourist's decision to visit Cuba, Vietnam, China or Saudi Arabia is not necessarily an endorsement of communism, a one-party state or a theocratic kingdom, but is more likely to be influenced by the perception that the destination is relatively stable and safe. Many of the world's 1.4 billion international tourists in 2019 were equally comfortable visiting democracies, dictatorships, monarchies, theocracies, nations under military rule and communist one-party states. The overwhelming proviso is that the political system and the governance apparatus supporting it are perceived as stable

In the 1990s, China introduced a set of criteria for outbound international travellers known as the Approved Destination Status (ADS) Agreements. The Chinese government applied a range of criteria as core conditions to permit Chinese travellers to visit specific destinations. Among the many criteria cited in ADS agreements was that the approved destination country was deemed by the Chinese government to be politically stable. By 2018, China had signed ADS agreements with 147 countries. By 2019, China accounted for ten per cent of global outbound tourism; consequently a share of China's outbound tourism market was sought by most countries (Arita et al., 2011).

A quick view of the world's top ten most visited countries reveals that a government that espouses liberal democratic principles does not in itself guarantee that it is a sought-after tourism destination. Of the top ten most visited countries in the world (Figure 4.1), most are multi-party parliamentary democracies but three, China, Thailand and Turkey, are subject to either military rule (Thailand) or have a regime dominated by a single political party (China). Although Turkey does have opposition parties and elections, political opposition to Turkey's ruling regime is severely restricted.

Government travel advisories are generally determined and disseminated by the foreign ministries of tourism generating countries around the world. They play a significant role in setting benchmarks, for determining whether a country is deemed politically stable or otherwise. There is a strong link between stability and perceived safety for tourists.

Rank and visitor numbers (millions)	Country	Form of government
1. 82.6	France	Multi-party parliamentary democracy
2. 75.6	United States	Multi-party parliamentary democracy
3. 75.6	Spain	Multi-party parliamentary democracy
4. 59.6	China	One-party communist
5. 52.4	Italy	Multi-party parliamentary democracy
6. 35.8	United Kingdom	Constitutional monarchy and multi-party parliamentary democracy
7. 35.5	Germany	Multi-party democracy
8. 35.0	Mexico	Multi-party democracy
9. 32.6	Thailand	Constitutional monarchy/military regime
10. 30.0	Turkey	Constitutional democracy/restricted autocracy

Figure 4.1 Ten most visited countries in the world (2017) and their political systems

Source: UNTWO, 2018. Adapted by the author

The wording of government travel advisories influences travel insurance coverage, airline routings, cruise destinations and wholesale tourism product (among many other elements) of international travel transactions in tourism generating countries. The development of the internet, apps and social media has facilitated the widespread online dissemination of government travel advisories within each tourism generating country. Both traditional and social media take their cues of destination safety from government travel advisories. Despite a high level of scepticism about the pronouncement of governments on many issues, the core role of government as a protector of its citizens tends to engender travel advisories with a high level of public legitimacy.

In a few countries, the travel industry plays a collaborative and consultative role in the shaping of travel advisories. In Australia, the Consular Consultative Group is largely composed of tourism-related stakeholder associations, which advise the Australian Department of Foreign Affairs and Trade on travel advisories and related tourism safety issues. This was the world's first such body, established in 2003. At the time of writing, only four other countries established similar advisory bodies: UK, Canada, New Zealand and the USA. It is relevant to this discussion that Australia, USA, UK, New Zealand and Canada share military and security intelligence under the provision of the 'five eyes' agreement. Although transnational and global tourism associations have advocated active consultation between the travel industry and national governments on the issue of travel advisories and tourism safety, most governments have been reluctant to engage in consultation with travel industry leaders on such issues. The main reason posited for this by some government officials is that security and safety are integral to the government's role, and consequently, should be the sole prerogative of government. The security of citizens is not regarded by most governments as a matter for negotiation, or in consultation with other parties (Beirman, 2019).

A core factor in the security rating of countries for the purpose of travel advisories is the perceived stability of a national government and its capacity to maintain effective governance over its territory. This allows for the peaceful transfer of power when a democratic country elects a government of a different political persuasion from the previous ruling party or governing coalition. Generally, when this occurs, while policies and political personnel will change, the fundamental administrative and security apparatus tends to be retained.

Until the 2020 Covid-19 pandemic, it was notable that the majority of countries for which Australian travel advisories have advised its citizens 'do not travel' were countries with high levels of political instability and internecine civil conflict; to put it more bluntly, failed states. Rarely was either the ideology or policy agenda of a government a significant consideration for a 'do not travel' advisory. Exceptions involved rare cases when bilateral relations degenerated into the severance of diplomatic relation or a state of war. Australia, New Zealand, Canada, the USA and UK engage in close cooperation on travel advisories and security assessments under the 'five eyes' agreement between the so called 'Anglosphere' countries in which sharing of security intelligence plays an important part.

For the 'five eyes' countries, security assessments of tourism for a given destination country rely on assessments from diplomatic legations located in the country. These assessments include the observations of expatriate business people operating in that country. A significant source is intelligence assessments, from either of the 'five eyes' states or allied 'reliable' intelligence sources. Occasionally, media and academic sources are considered (Kemish and Roach, 2006). Foreign ministries have traditionally been reluctant to consider the opinions of travel professionals in target countries even though tour operators, which specialise in specific destinations, often have a more intimate knowledge of those destinations than some of the more commonly accepted sources used by diplomatic legations. In Australia, this sceptical view on the part of government towards the travel industry has mellowed due to 18 years of predominantly harmonious cooperation between the Australian travel industry leadership and the Australian Department of Foreign Affairs and Trade on the issue of government travel advisories (Beirman, 2019).

Most government travel advisories urge tourists to purchase travel insurance prior to departure. Travel insurance generally covers cancellation, medical treatment for illness and accidents, and loss of baggage and valuables. Very few travel insurance companies cover policy-holders' loss for injuries or loss incurred as a result of politically motivated violence. In fact, most government travel advisories warn travellers to avoid being involved or visiting in the vicinity of any political demonstrations. Occasionally, this is unavoidable when demonstrations take place in airports, hotels or public places popular with visitors. The risk of injury and loss from political violence is magnified when travellers visit politically unstable destinations.

As discussed extensively in Chapter 3, the Covid-19 pandemic was a major disruptive influence on the shaping of government travel advisories. A pandemic and its public health implications, as opposed to issues of political stability, governed the safety and security assessment of virtually all countries in the world. Figure 4.2 illustrates this starkly. In 2020 due to the global spread of Covid-19, Australians, for the first time since Federation in 1901, were advised by their government to avoid all travel outside of their home country. Initially, contrary to some media reports, this was not a travel ban but was clearly a massive disincentive for Australians to travel internationally. By 25 March 2020, this advisory was upgraded to an international travel ban, apart from a small number of exemptions. The blanket Australian Covid-19 outbound travel bans exceeded the blanket 'do not travel' advisories applied by New Zealand and several other countries. Globally, most other countries imposed targeted Covid-19 travel bans based on perceived pandemic risk. These bans were largely unrelated to the issue of political stability or lack thereof. However, in some cases they were perceived as politically motivated measures. China has frequently accused the US government of imposing Covid-19 related travel bans for political rather than health-based reasons, especially after the Wuhan lockdown ended in late April 2020.

Incidents of crime, terrorism, health threats and natural disasters may lead a travel generating country to change the warning level of its travel advisory applying to the country in which this occurred. However, it is more likely to impact on a travel advisory applying to a country deemed to be politically unstable than for a country that is deemed as stable. In 2016 and 2017, there were a number of terrorist incidents in the UK targeting London and Manchester in which tourists were victims. However, these incidents had negligible impact on the travel advisories from the UK's major tourism source market countries. This was due to the widespread perception that the UK's political and security systems were stable and reliable. The terrorist attacks in London and Manchester in 2016 and 2017 had a minor impact on international tourism arrivals to the UK.

By contrast, a spate of terrorist attacks in Turkey from 2015 to 2017, some of which deliberately targeted tourists, led to a steep decline of international tourist arrivals during those years and the travel advisories of many of Turkey's key European source markets (notably Germany) were highly cautionary. In association with terrorist attacks, the Turkish government was subject to a coup attempt in July 2016, which almost overthrew the Erdogan government. Although the Turkish government re-asserted control over the country, political opposition in Turkey since the abortive coup attempt was repressed. Turkey's borders with Syria, Iraq and to a lesser extent Iran are not secure and Turkey has been engaged in a longstanding conflict with Kurdish separatists. These issues, in combination, led to a precipitous decline in tourism visitation to Turkey between 2015 and 2017. Turkey's tourism challenges have been to restore its negative image and its position as a top ten global tourism destination. By 2018, this recovery was clearly in evidence (APEC, 2017).

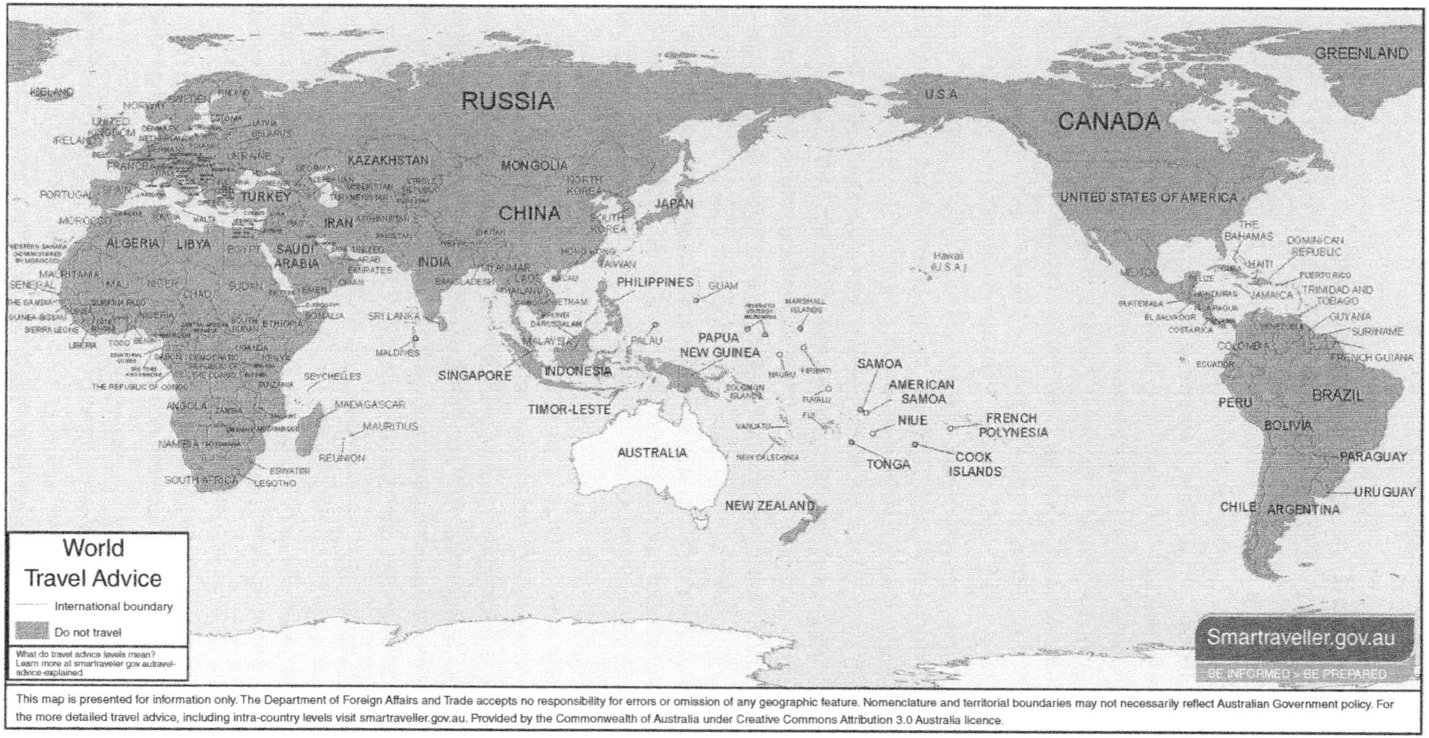

Figure 4.2 Australian government travel advisories in March 2020. This became a travel ban from 25 March 2020.

Source: Australian Department of Foreign Affairs and Trade, 2020. Provided under the Creative Commons 3.0 license.

Case Study 4.1

Political Instability and Tourism in the Middle East, 2010–2020

Political instability in a number of Arab countries since 2010 has had a significant impact on international tourism arrivals to formerly popular tourism destinations, including Egypt, Tunisia, Libya, Syria, Lebanon and Yemen. In all of these countries, regime change and political upheaval were accompanied by terrorist attacks and threats against tourists. The most extreme case of linkage between political instability and a tourism slump in the Middle East since 2011 was Syria.

Prior to the Arab uprisings (referred to by some optimistically naïve observers as the 'Arab Spring'), which commenced in Tunisia at the end of 2010, Syria was one of the success stories of Middle East tourism. Between 1996 and 2010, inbound tourism arrivals to Syria grew from just under one million international visitors in 1996 to 8.5 million in 2010 (World Bank, 2014). However, the beginning of anti-regime agitation in Syria in mid-2011 and the subsequent outbreak of a multi-factional civil war since that time has resulted in a precipitous decline in tourism, especially since 2012. Since 2012, Syria has been dominated by internecine conflict between Bashar Assad's regime and its supporters, which include Iran, the Iranian-backed Lebanese Shi'ite Hezbollah militia and Russia, who are fighting a disparate range of anti-regime groups, which have included Kurdish militias, Sunni Islamist groups and Islamic State (IS), coupled with strategic interventions by Turkey, the USA and occasionally Israel. This toxic, complex and brutal conflict has claimed the lives of over 500,000 Syrians and has led to the internal displacement of a further six million people and the emigration of over four million Syrian refugees to Lebanon, Jordan, Turkey and much of Europe.

In addition to the severe human cost of the Syrian civil war, many of Syria's most compelling and popular tourist attractions suffered severe damage during the conflict. The historic city of Aleppo and its extensive markets was largely destroyed (see Figure 4.3). During the fighting, Syria's most famous archaeological site, Palmyra, was looted and subject to extensive vandalism during the brief time of its capture and occupation by IS in 2015 and 2016. IS engaged in an active campaign to destroy non-Islamic archaeological artefacts in many areas, which have come under its control.

Despite an almost surreal attempt by the Assad regime to promote tourism to those few regions in the country unaffected by the civil war since 2016, tourism numbers to Syria can at best be measured in the tens of thousands. The conflicting agendas of the political forces involved in Syria's civil conflict means it is unlikely that the level of political stability which can support a significant tourism recovery will occur in the near future. This is despite the clear indications by mid-2019 that the Assad regime and its allies were militarily stronger than the anti-regime forces. By 2020, the Assad and his allies exercised military control over most of Syria. By early 2015, IS had imposed territorial control over ten per cent of Syria's land area. However, all IS territorial gains, including its 'capital city' of Raqqa (in eastern Syria), were conquered by all other parties to the Syrian civil war in 2019. IS remains a non-state terrorist threat to the Syrian regime and to neighbouring Iraq.

Figure 4.3 Destruction in Aleppo, Syria

Source: Image courtesy of Freedom House via Flickr. Shared under the CC BY 2.0 license.

Syria, represents the most extreme example of the political quagmire that has affected much of the Arab world since 2010, but Libya, Iraq and Yemen have also experienced a comparable level of political and social unrest, and politically-motivated violence.

Egypt has been the traditional tourism powerhouse of the Middle East/North African region. Egypt's archaeological attractions, dominated by the Pyramids of Giza, have been its traditional tourism drawcard. In 2010, Egypt attracted a record 14 million tourists. The ousting of Egypt's long-serving, autocratic president, Hosni Mubarak, in February 2011 followed weeks of popular protest in Cairo. Mubarak's overthrow was tacitly supported by Egypt's military. There was expectation among Egypt's intelligentsia that the demise of Mubarak's government would be followed by a democratic renaissance in Egypt. In fact, it was this hope (following the overthrow of the Tunisian regime just two months before) which gave rise to the term 'Arab Spring'. Despite the hope of the many young, politically idealistic, tech savvy, middle class Egyptians who participated in the protests, the aspirational democracy rapidly became an opportunity for a far better organised, conservative, Islamist political movement (the Muslim Brotherhood) to fill the political vacuum left by Mubarak's resignation.

From a tourism perspective, the timing of Egypt's uprising in February 2011 occurred during the peak tourist season and tourists in the country sought to depart Egypt rather than stay. Over 30 governments arranged evacuation flights from Egypt for

(Continued)

their citizens The Egyptian Tourism Authority (ETA) and local tourism industry sought to depict the overthrow of Mubarak as a new beginning for Egypt. The ETA largely ignored the real and perceived security threats to tourists. By the end of 2011, the hope of a democratic flowering in Egypt evaporated and the overthrow of Mubarak led to a longstanding and well organised Islamist political movement, the Muslim Brotherhood, filling the political vacuum.

The Muslim Brotherhood government in Egypt, led by Mohamed Morsi, came to power after winning elections in 2012. It was ousted by a military takeover in 2013. The military, led by Abdel Fattah el-Sisi, has effectively ruled Egypt since 2013. Although there have been a number of terrorist attacks in Egypt (notably in Sinai from 2013 to 2019), there has been a gradual recovery of tourism to Egypt.

A disturbing aspect of Egypt's attempt to lure tourists back was the myopic approach of the ETA. Rather than adopt the strategy they successfully employed in the 1990s, which directly addressed safety and security concerns (Beirman, 2003), the ETA, immediately following Mubarak's ousting in early 2011, ran a campaign that essentially led with a message, 'Egypt is now a democracy, we need money so come and visit.' Not only was the initial claim patently false: the refreshingly honest 'we need money' part of the campaign gave no assurance about how visitors to Egypt spending their money could be guaranteed a safe visit. In fact, from 2011 until the present day, there has been little beyond rhetoric to assure visitors to Egypt and to address their safety concerns, despite the approaching centenary of the discovery of King Tutankhamun's tomb in 2022.

Episodes of political instability in some parts of the Middle East had a ripple effect on tourism in countries not directly involved in civil unrest related to anti-regime uprisings. Jordan was significantly affected by the civil conflicts in neighbouring Syria and to a lesser extent, Iraq. Over 500,000 Syrian refugees have sought refuge in Jordan since 2011. However, Jordan's political stability was largely maintained despite a few outbreaks of politically motivated violence. In comparison to the political conflicts occurring in its neighbouring countries, Syria, Iraq and Israel, Jordan was an island of relative stability. Yet, between 2011 and 2016, there was a notable decline in Western tourism to Jordan based on a false but pervasive perception that Jordan shared the political and security problems of its neighbours (Beirman and Walters, 2017).

Conversely, the United Arab Emirates, especially Dubai and Abu Dhabi, have been highly successful in perceptually positioning themselves as sufficiently removed from conflicts in other parts of the Middle East to maintain steady tourism growth since 2011 (Thani and Heenan, 2017).

Political instability and its negative impact on tourism is not limited to the Middle East and North Africa. It is not within the scope of this book to discuss every country in which episodes of political stability have impacted on tourism. To do so would make for a very thick volume. Many countries experience outbreaks of political unrest, which may result from political movements attempting to violently assert their cause. This was the case in Thailand's yellow shirt versus red shirt conflict in 2009–2010 or civil unrest in Greece in response to EU economic austerity measures between 2015 and 2018. In both these cases, the more violent outbreaks of protests were confined to capital cities. However, extensive media coverage created a perception that unrest was nationwide.

It would be remiss in this chapter not to discuss a very recent and very dramatic example of a destination that was long an iconic tourism destination and now faces a significant reputational challenge brought on by political instability – Hong Kong. The story of Hong Kong's anti-government protest movement, especially since 2019, includes a number of sub-plots, including the growing 'great power' international assertiveness of China, it's conflict with the USA and the Chinese Communist Party's rejection of the 'one country, two systems' approach. The current Chinese regime, led by Xi Jinping, has sought to limit cultural, social, religious and political pluralism in China. Movements and ideas deemed to challenge or threaten the dominance of the Communist Party have been subject to repression.

Case Study 4.2

Hong Kong's 2019–2020 Protests and Tourism

In 2019, one of the most dramatic episodes of political instability impacting on tourism erupted in a destination which for over a century has based its wealth on trade and tourism – Hong Kong. When China regained sovereignty over Hong Kong from Britain in 1997, Hong Kong was initially granted a high degree of political and administrative autonomy by the Chinese government, under China's 'one country, two systems' policy. In fact, a core condition of the handover was that Hong Kong could exercise administrative autonomy for 50 years after the handover year of 1997. Hong Kong retained its own elected government administration although the Chinese central government exercised a high level of influence.

From a tourism perspective, international visitors to Hong Kong, including Chinese citizens from the mainland, were treated as foreign visitors and required to undergo customs and immigration procedures. Hong Kong's economy enjoyed a high degree of independence and maintained its own police and security infrastructure. Hong Kong still has its own currency, the Hong Kong dollar, which was convertible with the Chinese RMB (renminbi).

In practice, China exercised a high level of political control over Hong Kong's political leadership. Hong Kong's autonomous status has made it a very popular

(Continued)

destination for international leisure and tourists, and especially tourists from mainland China who were attracted to the shopping and trade. In 2018, Hong Kong hosted 65.14 million international visitors, of whom just over 51 million were from mainland China and 14.1 million from the rest of the world. International arrival figures have been in significant decline (see Figure 4.4) since the protests began in June 2019 (Hong Kong Tourism Board, 2020). Hong Kong has traditionally based its economy on trade and tourism, both of which are under severe threat to date.

There were long-standing concerns in Hong Kong that China's exercise of political control compromised the autonomy of Hong Kong. In June 2019, demonstrations advocating greater freedom of expression in Hong Kong intensified when the Hong Kong government threatened to extradite demonstrators for trial to mainland China. Although Hong Kong's government, with the approval of Beijing, backed down from this policy in July 2019, mass demonstrations demanding political independence continued to intensify.

By November 2019, the demonstrations, especially at Hong Kong's universities escalated into violent clashes between Hong Kong police and demonstrators. The real concern for Hong Kong residents is the possibility that the central Chinese government would intervene and violently crush the demonstrations, and oblige Hong Kong residents to be subject to Chinese legal, social, economic and political control. At the time of writing, the Chinese central government has demonstrated a mixture of profound irritation and military restraint as it became concerned about the regional and global political and economic reaction to crushing a political insurgency in Hong Kong. The hostile international response to China's 1989 crushing of political protest in Beijing is fresh in the mind of the Chinese government.

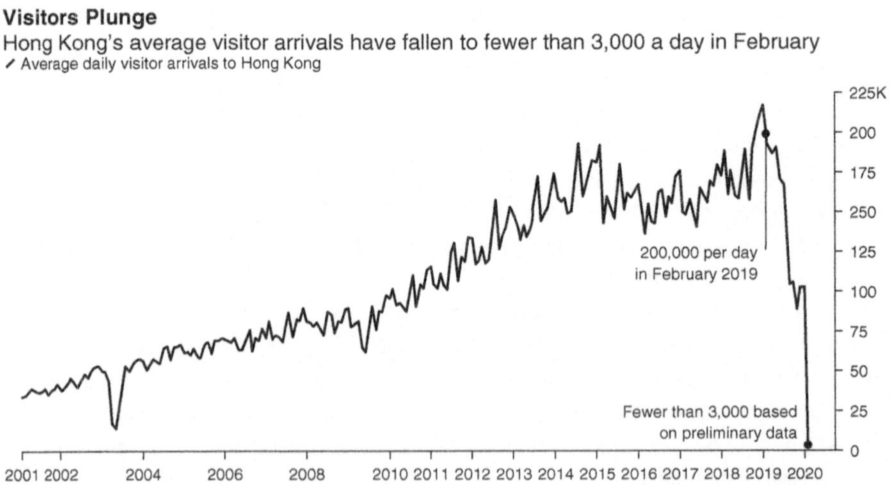

Figure 4.4 Hong Kong tourism arrivals, 2001–February 2020 (thousands)

Source: Hong Kong Tourism Board, 2020. Used with permission of Bloomberg LP. Permissions Copyright © 2021. All rights reserved.

As a traditional *entrepôrt* country, Hong Kong's attraction for both business and leisure visitors has been its relative freedom in comparison with China. Traditionally, China has regarded Hong Kong's autonomy as a strategic facilitator of trade between between China and the rest of the world. However, from China's perspective, unchecked political unrest on China's doorstep created a perception, internationally and within mainland China, of a challenge to the legitimacy and security of China's ruling regime. In the Chinese context, this would signal a global loss of face.

As these tensions grew, both business and tourism were under severe threat. Hong Kong has long been a popular venue for conferences and conventions and political instability is toxic to hosts, conference/convention organisers and prospective international attendees. Media images of violent demonstrations near major hotels, the international airport and many popular tourism attractions deterred both leisure and business tourists. Hong Kong's shopping attractions attracted far fewer customers, and as instability continued, many high-end retailers and large businesses closed or downscaled their presence in Hong Kong.

From a tourism perspective, the response of the HKTB to the political protests was akin to watching the three wise monkeys in action (hear no evil, see no evil, speak no evil). If one was to believe the HKTB's website, all was well in Hong Kong. On rare occasions, the HKTB's site would mention 'processions' disrupting passengers at Hong Kong's airport, blockading a major hotel or impacting on public transport. Apart from those occasional coy observations, the HKTB would be the last place a prospective tourist or in-country visitor would visit when seeking information on the impact of political unrest on tourism. Ironically, prospective and in-country tourists could learn more about the political unrest in Hong Kong from TV news reports than from the HKTB.

As an arm of government, in the HKTB of Hong Kong and in its international offices, public communication was subject to strict control by both the Hong Kong government and the central Chinese government. This included avoidance of disseminating bad news, even if by doing so they were exercising a duty of care to tourists. Ironically, during the SARS crisis of 2002, the HKTB enjoyed a great deal of freedom to market Hong Kong back to recovery and enable all its stakeholders to be fully informed about the facts on the ground. The evidence of 2019–2020 indicates that the HKTB no longer enjoys that freedom.

During 2020, the heightened threat of Covid-19 effectively reduced international tourism to Hong Kong tourism to a trickle. The Chinese central government and the Hong Kong government banned tourism between mainland China and Hong Kong from the Chinese New Year, mid-January 2020 until the end of April 2020. Since March 2020, China has threatened to arrest and extradite Hong Kong citizens engaged in political activity deemed as hostile to the peace and security of Hong Kong to the mainland. While these measures have not stopped anti-Beijing demonstrations, they have clearly reduced their frequency and intensity.

(Continued)

The ongoing political instability in Hong Kong, exacerbated by the impact of Covid-19, appears to signal the end of China's commitment to 'one country, two systems', which was the basis of Hong Kong's longstanding appeal as both a global *entrepôrt* and a tourism destination. The HKTB appears so effectively muzzled by the Chinese government that it is no longer in the position to freely operate a credible recovery campaign that could be trusted by its foreign tourism stakeholders.

Any moves to restore tourism to Hong Kong or resolve the political impasse between the Chinese government and Hong Kong independence supporters were overridden by the impact of Covid-19 during 2020. By mutual agreement between the Hong Kong authorities and the central government in Beijing, tourism between the mainland and Hong Kong ceased during most of 2020. Although the virtual closure of Hong Kong reduced Covid-19 to manageable proportions, the future of tourism to Hong Kong awaits a political resolution.

A Broader Profile of Political and Policy Instability, and its Impact on Tourism

There is far more to political instability than riots, civil war and violence. Major changes in government policy may not necessarily lead to riots in the streets, but they can arouse significant uncertainty for tourism. Radical changes in policy when governments change hands have the potential to have a negative impact on tourism. In Kenya, an outbreak of political violence during and after the 2007 elections led to months of political instability, which led many visitors to abandon Kenya as a tourism destination after four years of spectacular growth between 2003 and 2007.

Although Thailand has been a global role model of tourism resilience since 2000, the military coup that installed a military government in 2014 led to a significant decline in tourism during that year.

On a more dramatic level, the collapse of communism in Europe and the former Soviet Union during the late 1980s opened up many formerly closed destinations to tourism. While this was a welcome opportunity for tourism in these countries, the political adaptation many former Soviet satellite countries experienced, including Hungary, Poland, Romania, Slovakia, the Czech Republic and Albania, was economically and socially disruptive. East Germany, which endured decades of brutal and invasive totalitarian rule until 1989, was incorporated into a re-unified democratic Germany after the Berlin wall was torn down.

The most violent chapter after the collapse of communism involved the breakup of the former nation of Yugoslavia into states based on ethnic enclaves. The early 1990s was dominated by wars involving Serbia, Croatia, Bosnia-Herzegovina and Kosovo. These violent conflicts severely impacted on tourism, especially to Croatia's traditionally popular Adriatic Coast and islands (Beirman, 2003). Of the former Yugoslavian republics, only Slovenia and Montenegro were fortunate enough to avoid conflict.

Central to the recovery of tourism in Eastern Europe was the appeal to tourists being able to visit destinations, which were previously off limits or exceedingly difficult to visit. For countries such as the Czech Republic, its historic capital Prague became a mecca for tourists from all over Europe and the wider world from the 1990s. Following the collapse of communism in Eastern Europe in the late 1980s and early 1990s there was a commensurate growth of international tourism to Slovakia, Hungary, Romania, Poland, Bulgaria, Albania, Russia and many newly independent former Soviet republics (including Belarus, Ukraine, Lithuania, Latvia, Estonia and Armenia to name a few) as these formerly closed off countries opened their borders to tourism.

Case Study 4.3

Brexit – Policy Confusion Creating Uncertainty for Tourism

In 2016, the British government, then led by Conservative Party Prime Minister David Cameron, held a referendum on the issue of Britain's membership of the European Union (EU). There was a sharp division of opinion in the UK on the issue of either retaining the UK's affiliation with the EU (the Remain camp) or severing its affiliation with and membership of the EU (the Brexit or Leave camp). The issue of Britain's membership had been a running sore in UK politics since 1973, when the UK became a member of the EU during the prime ministership of Edward Heath (Conservative). Ironically, much of the debate between pro-Europeans and 'Eurosceptics' was conducted within the ranks of the conservative elements of British politics (Cameron, 2019).

From a tourism and trade perspective, Britain's membership of the EU had many benefits. Companies offering goods and services could trade with zero tariffs throughout the EU, essentially increasing their potential markets by 500 per cent.

UK citizens were free to travel and work in any of the EU countries and EU citizens were entitled to travel and work in the UK. Although the UK maintained border controls for entry into the UK, and its pound sterling currency, EU citizens could travel to and stay in the UK with few restrictions and Britons could travel to and between EU countries with minimal formalities. This led to a surge in travel to the UK from continental Europe and a corresponding surge of British tourism to EU countries. The massive growth of the intra-European market enabled travellers to experience high quality travel experiences at very low prices. Many British travellers found it cheaper to have a two-week holiday in France, Italy, Spain or Greece than a holiday in the UK.

Many middle class and wealthy British took advantage of the opportunity to purchase property in Europe, which was generally less expensive than comparable

(Continued)

property in the UK. In some cases, parts of these properties were transformed into luxury B&B businesses for semi-retired British who could enjoy the sun, earn some local income and pay lower taxes than at home.

There was a surge in competition among the airlines operating services between the UK and European cities. The addition of low-cost carriers such as easyJet, Ryanair, Norwegian Air Shuttle, Wizz Air and Eurowings led air fares between the UK and EU countries to plummet between 1973 and 2019. The opening of the Eurostar high-speed rail route in 1994 from London to Paris, Amsterdam or Brussels using the Channel Tunnel between Dover and Calais gave travellers between the UK and Europe the choice of ferry, rail or airline transport.

While the benefits of the UK's EU membership were undeniable from a tourism perspective, many Britons chafed at the imposition of EU directives from Brussels, which set controls over agricultural production, manufacturing standards and a range of day-to-day activities that many Briton's did not believe was the business of the bureaucrats of Brussels. Some British were concerned about what they perceived as EU citizens flooding into Britain, accepting low wages and taking 'British jobs'. Some more conservative British baulked at their traditional navy-blue British passport being printed with a red cover, renamed as a European passport and full of details in German, French and other European languages.

Centuries as an island nation led to concerns among more conservative British that Britain's distinctiveness from 'the continent' was undermined by its EU membership. Fundamentally, many British citizens resented that the UK's entry to the EU in 1973 resulted from an act of parliament in 1972, and until the 2016 referendum, Britain's original EU membership was never put to a national vote.

On 23 June 2016, after weeks of acrimonious public and media debate, the UK's Brexit referendum was held. The result was very close. Just under 52 per cent of British voters chose to leave the EU and just over 48 per cent chose to remain in the EU.

The referendum result confounded the majority of both British and European political pundits. Most of them thought the arguments in favour of the Remain case, advocated by David Cameron's government and much of the British and European 'commentariate', would easily prevail over the Leave arguments. The Leave case's leading supporters were the UK's current Conservative prime minister, Boris Johnson, and Nigel Farage, leader of the Brexit Party and ironically (given his party's core policies) a European member of parliament. Notably, the opposition Labour Party, led by Jeremy Corbyn, maintained a neutral posture during the Brexit debate.

In essence, the referendum vote was the simple piece of the Brexit puzzle. Neither the EU nor the British government developed a contingency plan addressing the practical implications of Brexit. This opened a Pandora's box of confusion, questions and doubt. Prime Minister David Cameron resigned shortly after the referendum, having staked his political reputation on a Remain vote. He was replaced as Conservative leader and prime minister by Theresa May. May's Prime Ministership (2016–2019) was dominated by acrimonious and inconclusive negotiations with the EU over the terms of the UK's departure from the EU. While May's government was committed to implementing the electorate's decision, many cabinet members were less than enthusiastic about Brexit. The absence of an agreement between the UK and the EU delayed the break between Britain and the EU. The full terms of Brexit agreement were fully resolved at the end of December 2000 and the free movement of travellers included in the Brexit agreement between the UK and EU was muddied by the border restrictions imposed in reponse to Covid-19.

From a risk management perspective, Brexit has been an object lesson of how not to manage risk. The EU never made a serious attempt to establish a set of protocols in the event of a country choosing to leave it, and when David Cameron made the decision to hold the referendum in 2016, his government hadn't developed a contingency plan in the event of either a remain vote or a leave vote (Cameron, 2019).

Dissatisfaction within the Conservative Party over the delays with Brexit's implementation led to Theresa May's ousting as leader in June 2019. Her replacement was Boris Johnson, an outspoken proponent of Brexit. Johnson led the Conservative Party to a massive win in the December 2019 general elections. Although Johnson claimed his win was an electoral mandate for Brexit, the extent of the Tory victory was arguably more a repudiation of the Labour Party leader Jeremy Corbyn's Marxist economic agenda than an endorsement of Brexit. Corbyn's attitude to Brexit remained a mystery to most of the British electorate before, during and after the referendum.

Although Boris Johnson promised a rapid resolution of the Brexit impasse with Europe, negotiations on this issue during 2020 were overshadowed by the global challenge of Covid-19. The pandemic has been damaging for many European countries, especially Britain. By October 2020, Britain had the highest European death toll from Covid-19 (over 900,000 cases and over 46,000 deaths). The impasse between the UK and EU appears likely to lead to a 'no deal' Brexit (no agreement). In turn, this leaves multiple unresolved issues, especially those applying to tourism movements and visas.

The implication of Brexit on two-way tourism between the UK and EU countries are significant. In 2017, of the top ten source national markets to the UK, seven were EU countries (Tourism Alliance, 2017). They include France, Germany, Spain, Netherlands, Italy, Republic of Ireland and Belgium. The only non-EU countries in the UK's top ten source markets were US (ranked first), Australia (ranked ninth) and Sweden (ranked tenth).

Conversely, British travellers visited EU destinations in their millions with sun, sand and sea destinations such as Italy, Greece, Spain, France, Croatia, Cyprus and Portugal being especially popular.

(Continued)

The unresolved tourism-related issues of Brexit included:

1. Freedom of movement allowing EU citizens the right to live and work in the UK, and granting reciprocal rights for British people to live and work in the EU. This has been a powerful motivator for two-way tourism between the UK and EU countries, and there is no certainty that this right will remain once Brexit is finalised. The entire issue of visas and attached conditions, restricted stays etc. is unresolved.
2. A related issue is the bilateral right to work. From a tourism industry perspective, between 12 and 23 per cent of tourist sector workers in the UK are EU nationals. If the bilateral right to work is removed, this creates significant uncertainty about the rights of UK citizens to work in Europe and for Europeans to work in the UK. From a tourist perspective, many young travellers engaged in a gap year could literally fund their travel by taking on jobs in EU countries. There is no longer certainty that this right will remain for bilateral travel between the UK and Europe.
3. Value of the pound in relation to the euro. Most but not all EU countries adopted the Euro as their official currency in 2002. With uncertainty surrounding both tourism and trade (especially free trade provisions), the UK pound has lost value against most hard currencies since the 2016 referendum.
4. The free trade association of the EU meant that tariffs between EU members were abolished. With Brexit, the issue of tariffs between Britain and the EU remains unresolved and this may significantly affect the cost of goods, services and trade costs between the UK and EU countries.
5. Aviation agreements have become a vexed Brexit issue. The EU abolished all bilateral airline agreements applying to the EU (including Britain) and the EU was subject to an open skies' agreement. There is no certainty that this will continue post-Brexit or whether restrictive bilateral agreements may apply between British carriers and EU countries post-Brexit.
6. Will British citizens enjoy the same rights for owning property, especially if they use them for business purposes in EU countries, post-Brexit, as they did pre-Brexit?
7. Will Brexit have an impact on duty free rights and allowances for travellers between the UK and EU countries?

As Lim (2019) points out, the uncertainty surrounding the tourism implications of Brexit aroused speculation, which in turn led to swings in demand for tourism between the UK and EU countries.

In the four years since the Brexit referendum in 2016, the questions surrounding the impact of Brexit on tourism and trade remained unresolved until the end of December 2020. While this may not be a question of the stability of the British government, a major change affecting the UK's trade and tourism relations with the EU is almost certain to have a major impact (in all likelihood negative) on tourism between the UK and EU. Until the terms of Brexit are clearly defined, it's impossible to predict the statistical impact of Brexit on tourism, if any.

Postscript on Brexit, January 2021

On 31 December 2020, an agreement between the UK and EU on the conditions of the UK's separation from the EU was finally signed, literally hours before the deadline. As expected, the rights of two-way tourism between the UK and Europe have been curtailed. Although there is relatively open two-way leisure tourism between the UK and EU, the rights of unrestricted employment between the UK and Europe have been severely curtailed. Full details of the agreement and its terms and conditions are available from the EU and the UK government (European Commission, 2021).

RECOVERY STRATEGIES FOR DESTINATIONS AND BUSINESSES AFFECTED BY POLITICAL INSTABILITY

As the examples and the three case studies above demonstrate, there is no one strategy applicable to the recovery of a tourism destination that has experienced episodes of political or policy instability.

Much depends on the extent and degree of the instability, its impact on access to the destination, tourism infrastructure and professional capacity to deliver a tourism service. Critically, the perception of source markets plays a major role in the nature and speed of destination and tourism business recovery.

Based on the extreme example of Syria, which has experienced widespread devastation, casualties and emigration, the restoration of Syria as a desirable tourism destination is likely to take many years. The rebuilding of basic infrastructure and urban areas could involve at least two years' work. A national recovery of tourism in Syria would need to rely on objective, independent assurances that a resumption of tourism would be safe for tourists and tourism industry professionals. As the Syrian civil war has involved multiple protagonists and left unused explosives, mines and improvised explosive devices (IEDs), it would take considerable time to clear the fields, cities and towns of the detritus of war. The critical question for Syria is whether the ruling regime really exerts political and security control over all of Syria.

By contrast, if the UK and Europe are able to come to a clear and unambiguous agreement on Brexit, which results in few substantive changes and causes minimal disruption to tourism, the impact of Brexit's uncertainty on tourism between the UK and Europe may be minimal. Conversely, if the outcome of Brexit involves a more restrictive set of entry requirements for tourists between the UK and Europe, this has the potential to negatively impact on demand for two-way tourism between the UK and Europe.

The situation between China and Hong Kong remains complex. The mainland Chinese government appears determined to impose its political control over Hong Kong. However, the independence faction in Hong Kong can leverage the right of Hong Kong citizens to leave Hong Kong and accept offers made by the UK, Australia and some other countries to take up citizenship there as leverage to embarrass the Chinese government, and to goad other countries and the international community into criticising China. Political instability at its heart is centred on the concept of uncertainty. As the Syrian and the Hong Kong case studies demonstrate, political uncertainty is not conducive to attracting international tourism.

In episodes of political and security uncertainty, tourists have some key concerns that need to be addressed from a reliable and trustworthy source. These concerns include:

- Access into the country and reliable egress out of the country.
- Safe internal transport.
- Accommodation opening and operating.
- Reliable electrical power and communications.
- A reliable banking and foreign exchange system.
- Major attractions being open and secure.
- A law and order regime subject to legitimate jurisdiction.
- A perception that a visit to the destination is safe.

Thailand has experienced numerous episodes of political instability in recent years. Yet, with few exceptions, international tourism rose annually between 2000 and 2019. During periods of civil unrest (notably 2009–2010), the Tourism Authority of Thailand's (TAT) website gave hourly updates on the status of transport operations, accommodation opening and access to attractions, shopping centres and other significant public buildings.

The TAT has generally been a trusted and reliable source of information for both intending tourists and tourists visiting Thailand. Timely and accurate information on the status of tourism in Thailand, even during times of political instability, was clearly communicated to all key tourism stakeholders. The building of trust is a long-term process, which is vital to destination management and marketing organisations. In instances of political unrest, it is vital for DMOs to make a distinction between problematic parts of the destination and the broader context that the destination is safe to visit.

Between 2008 and 2018, there was considerable political unrest in Greece over the EU's attempts to impose financial discipline over the explosion of Greek sovereign debts.

This led to bitter and widely reported protests centred on the parliamentary precinct of the Greek capital, Athens. However, outside Athens public dissatisfaction rarely manifested itself in the form of protests. The Greek National Tourism office sought, with mixed success, to make that distinction clear to its stakeholders and invited travel industry leaders and media to observe the distinction.

SUMMARY AND CONCLUSION

Political instability or the perception of political instability is a major risk for tourism destinations. The key concerns centre on:

- Stability of the government and governance.
- A perception and reality that the destination is safe to visit.
- Reliable sources of information.
- A degree of certainty that tourism services and administration operate reliably.
- Certainty that tourists can enter, exit and experience the country safely.

For most of the 21st century, safety has assumed a paramount concern for all tourists. A volatile political environment is not necessarily a danger to tourists if that volatility is lawful, non-violent and concentrated in specific areas. Conversely, a volatile political environment in which civil order has broken down is a very dangerous place for a tourist to visit. Political instability, especially when it manifests itself in the form of civil war, widespread unrest or a failed state, is toxic for tourism. Uncertainty and a lack of reliable and trustworthy information tends to deter tourists from visiting an affected destination.

In reference to the cases in this chapter, Syria is a failed state and a highly dangerous destination. Most major tourism attractions and destinations in Egypt are relatively safe but the ETA has proven to be an unreliable source of information since 2011 on safety and security matters. If and when a DMO provides inaccurate and unreliable information, tourists need to find alternative sources of reliable information.

In the case of Brexit, the longstanding policy uncertainty represented a risk in terms of the freedom to travel between the UK and Europe for over three years. The combination of a vague Brexit outcome and Covid-19 related internal restrictions within the UK, and the fluidity of the travel bubble between the UK and EU countries, make planning travel between the UK and Europe subject to uncertainty.

Uncertainty dominates the future of tourism to Hong Kong and much of this uncertainty is related to China's ultimate intentions towards the degree of Hong Kong's future autonomy. Associated with this is the acceptance or rejection of that degree by Hong Kong's citizens. In common with Egypt, a muzzled HKTA is an unreliable guide to the status of tourism in Hong Kong.

Political instability, in all its many forms, is a risk that every destination needs to factor into their risk and crisis management planning.

Discussion Questions

1. What are five key forms of political instability and how can they impact on tourism?
2. Discuss a strategic approach for DMOs to legitimately restore tourism confidence from key source markets in a destination that has been affected by an episode of political instability.
3. Can tourism be viable during episodes of political instability, and if so, how can DMOs in good faith maintain the promotion of tourism when parts of that destination are affected by episodes of political instability?
4. Government travel advisories are regarded by intending travellers as an important guide to destination safety. Discuss why this is the case.
5. How important is the role of DMOs in addressing safety concerns for travellers who visit countries experiencing episodes of political unrest? Discuss this in reference to at least two countries.
6. In relation to the Brexit agreement terms governing tourism between the UK and Europe, do you feel that it achieved the best outcome from mutual perspectives of the UK and EU?

REFERENCES

Arita S., Edmonds, C. and La Croix, S. (2011) 'Impact of approved destination status on Chinese travel abroad', *Tourism Economics*, 17(5): 983–96.

Asia-Pacific Economic Cooperation (APEC) (2017) *Strengthening Tourism Business Resilience Against the Impact of Terrorist Attack*. Report of the APEC Counter-Terrorism Working Group's Workshop on 9–10 May 2017, Bali, Indonesia. APEC Counter-Terrorism Working Group, APEC Secretariat: Singapore.

Australian Department of Foreign Affairs and Trade (2020) www.smartraveller.gov.au. Accessed 3 December 2020.

Beirman, D. (2003) *Restoring Tourism Destinations in Crisis. A Strategic Marketing Approach*. Allen & Unwin: Crows Nest (Sydney), Australia. pp. 227–42.

Beirman, D. (2019) 'Collaborative approaches to government travel advisories in Australia between Australia's tourism industry leadership and the Australian Department of Foreign Affairs and Trade', *Journal of Vacation Marketing*, 25(1): 71–87.

Beirman, D. and Walters, G. (2017) 'Marketing the Middle East in times of political instability – the case of Jordan', in Almuhrzi, H., Alryami, H. and Scott. N. (eds), *Tourism in the Arab World. An Industry Perspective*. Channel View Publications: Bristol, UK. pp. 175–87.

Cameron, D. (2019) *For the Record*. William Collins: London. (This book is a useful resource from the former British prime minister to help understand the political processes at play before, during and after the Brexit debate and the 2016 referendum.)

European Commission (2021) *EU, UK Withdrawal Agreement*. https://ec.europa.eu/info/relations-united-kingdom/eu-uk-withdrawal-agreement_en. Accessed 6 January 2021.

Gerschewski, J. (2013) 'The three pillars of stability, legitimation, and co-optation in autocratic regimes', *Democratization*, 20(1): 13–38.

Hong Kong Tourist Board (HKTB) (2020) www.discoverhongkong.com. This is a portal to any sites relevant to tourism in Hong Kong. Accessed 16 December 2020.

Kemish, I. and Roach, J. (2006) 'Bali bombings: A whole of government response', in Wilks, J., Pendergast, D. and Leggat, P. (eds), *Tourism in Turbulent Times: Towards Safe Experiences for Visitors*. Elsevier: Amsterdam. pp. 277–90.

Lim, R. (2019) *Economic Outlook of Brexit on Retail*. http://retaileconomics.co.uk/video-library/economic-outlook-brexit-retail-2019. Accessed 3 December 2020. Richard Lim is the CEO of Retail Economics UK and a frequent commentator on Brexit.

Thani, S. and Heenan, T. (2017) 'The UAE: A Disneyland in the desert', in Almuhrzi, H., Altiyami, H. and Scott, N. (eds), *Tourism in the Arab World*. Channel View Publications: Bristol, UK. pp. 104–116.

Tourism Alliance (2017) Downloads on UK Tourism. https://www.tourismalliance.com/downloads/TA_395_420.pdf. Accessed 5 May 2021.

UN World Tourism Organisation (UNWTO) (2018) *UNWTO Tourism Highlights. 2018 Edition*. UNWTO: Madrid, Spain.

Wahab, B. (2019) 'What really causes instability in the Middle East?', in Rubin, M. and Katulis, B. (eds), *Seven Pillars of What really Causes Instability in the Middle East*. AEI Press: Washington, USA.

World Bank (2014) *World Development Report: Risk and Opportunity. Managing Risk for Development*. World Bank: Washington, DC. pp. 53–75.

5

THEME 2: TOURISM RECOVERY FROM TERRORISM

INTRODUCTION

Until the Covid-19 pandemic outbreak of 2020, terrorism and its potential, actual and historical impact on tourism aroused more intense media coverage than any other category of risk to tourism.

One of the main reasons that terrorism attracted so much attention was the highly politicised and overt nature of terrorist violence. An act of terrorism involves a deliberate act designed to kill, maim and damage people and property in the interests of a political, ideological or religious cause accompanied by the intention of using and publicising violence as a means to espouse that cause.

There is a vast body of academic literature devoted to terrorism, its motivations, causes and consequences. There is also a vast body of political and philosophical literature that debates the very definition of terrorism as it applies to certain groups engaged in ideological, political or religiously motivated violence.

The old cliché that 'one man's terrorist is another man's freedom fighter' has as much resonance today as it did in the many post-colonial independence struggles, which occurred following World War II. Many famous and at times revered heads of state and heads of government were, for part of their life, labelled as terrorists. Nelson Mandela, Jawaharlal Nehru, Mao Tse Tung, Ho Chi Minh, Fidel Castro, Menachem Begin, Gamal Nasser, Jomo Kenyatta and Sukarno are just some of the heads of government/state in modern history who were described by their opponents as 'terrorists' during their political careers and their rise to political power. Most of them represented examples of traditional terrorism. These were often on the fringes of post-colonialist independence movements in Africa and Asia emerging after World War II. Many European colonial powers found their capacity to maintain colonial rule diminished. The occupation of their country or colonies by either Germany or Japan gave impetus to independence movements, which saw a window of opportunity to press their claims.

The definition of a terrorist is subject to political and ideological narratives. Political narratives promoted by partisans from either side of a political conflict generally described leaders and combatants supporting their cause as defenders and fighters for either freedom or a just cause. They labelled their opponents as terrorists, oppressors, colonialist, racist, morally bankrupt or variations on these themes. Integral to understanding the difficulty in achieving a resolution between conflicting visions of national self-determination in conflicts, such as the Israeli–Palestinian conflict, has been the tendency (especially by the more extreme protagonists) to dehumanise and delegitimise the 'other side's' national/religious and historical connections and aspirations. Delegitimisation and dehumanising the 'enemy' or the 'other' thus justifies politically motivated violence against them. Historically, many nationalist movements have sought to employ terrorism as a tactic to achieve or hasten national self-determination. In the European context, groups such as the Irish Republican Army (IRA), ETA (Basque Separatists in Spain) acted as separatist groups to advance a separate state carved out from a larger power. In the Middle East, Kurdish groups who live on the frontier of regions of Syria, Turkey, Iraq and Iran have employed terrorism to highlight their cause to create a distinctive Kurdish state.

Some extremist groups (notably Islamic State (IS)/*Daesh* and its supporters), which routinely targeted Muslims and non-Muslims, justified the murder of fellow Muslims who did not follow their particular Sunni, Salafist stream of Islam on the grounds that they were 'unbelievers'. IS in common with many terrorist and extremist groups dehumanised their enemies and opponents in a manner acceptable within their ideological framework to eradicate them. Jihadists would dismiss terrorist targets as *kuffars* (nonbelievers) just as Nazis and neo-Nazi groups dismissed non-Aryans as *untermenschen*, sub-humans.

From a media perspective, there is a mixture of headline producing horror ('if it bleeds it leads') and prurient fascination surrounding the motives of people who explode a suicide vest, killing themselves and dozens around them in the interests of a religious, ideological or political cause. Where terrorism and tourism tend to collide is due to tourists often been perceived as soft (vulnerable) and highly-visible targets for terrorist attacks. The killing, injuring or abducting of tourists from multiple nationalities provides an opportunity for terrorism's greatest ally and asset to come into play – widespread international publicity. Targeting tourists from multiple countries, irrespective of whether the attack takes place in Turkey, Burkina Faso, Thailand, Tunisia, Spain, Bali, France, Sri Lanka or the USA is certain to generate headlines in every country in which a victim resides, and terrorist attacks feature in every traditional and social media platform.

Peter Tarlow (2014) in his book *Tourism Security* explains that for terrorists, publicity and notoriety are essential in the promotion of a political, ideological or religious cause. While both criminals and terrorists may share a desire to rob, injure, abduct or kill their victims, the criminal seeks to do so anonymously and avoid public or media exposure. Conversely, terrorists and their sponsor organisations generally 'claim credit' for their actions and will issue traditional media and social media releases and statements, which seek to 'justify' and explain their actions and motivations. During prolonged terrorist actions such as hostage taking, the perpetrators often seek an opportunity to explain to an international and occasionally a global audience why they are engaged in this act. In fact, media access is frequently included in the negotiating strategy of terrorist groups with government officials, police and military forces at the site of the incident.

For some terrorist groups, notably Islamist jihadists, Western tourists engaged in secular leisure activities involving scant attire, provocative sexuality, music, dancing and consumption of alcohol and pork are deserving of the ultimate punishment for behaviour deemed offensive to Islam (as the perpetrators perceive it). This point of view positions the tourist as an enemy of Islam, whose behaviour and actions justify and even invite severe punishment.

TERRORISTS TARGETING TOURISTS AND TOURISM

There have been many instances around the world in which terrorists have deliberately chosen tourists as their primary target. At the time of writing, the deadliest terrorist attack specifically targeting tourists (since 2000) was that conducted in 2015 by supporters of IS who planted a bomb on a Russian civilian airline at Sharm El Sheikh, Egypt, which resulted in the deaths of over 224 Russian holiday makers and the crew of their Metrojet charter aircraft. The bomb was designed to explode when the aircraft reached a specific altitude. The subsequent explosion killed all passengers and crew.

The 12 October 2002 bombing of the Sari Club and Paddy's Bar in Denpasar, Bali, by Islamists from the group Jemaah Islamiyah, led to the deaths of 202 people and the wounding of over 400 people. This twin target attack killed and injured tourists from 27 countries, including 88 Australians killed and 200 wounded. This attack remains one of the largest mass casualty terrorist attacks targeting tourists in recent history (Gurtner, 2007).

Although there have been many terrorist attacks that have specifically targeted tourists in recent years, many more tourists have been victims of terrorism that involve random targeting. The IS terrorist attack in Paris of 13 November 2015 killed over 135 people at restaurants, the Bataclan concert venue and a sporting stadium, indiscriminately targeting anyone at those venues. Most of the victims were Parisians enjoying a night out and some were tourists. The 21 April 2019 terrorist attacks in Sri Lanka killed over 250 people of whom 39 were guests staying in three upmarket Colombo hotels, targeted by suicide bombers. The remaining 210 fatalities were worshippers in three Christian churches attending Easter Sunday services, targeted by suicide bombers allegedly acting in the name of IS.

Many of the world's most notorious terrorist attacks in recent years either included tourists among the victims or have affected the motivation of tourists to travel to the destination. The 9/11 attack on the World Trade Center in New York and the Pentagon (Washington, DC) in September 2001 killed airline passengers, office workers, visitors to the World Trade Center and military personnel and staff in the Pentagon. This attack killed almost 3,000 people of whom 372 were citizens of 62 countries other than the USA; some of the victims were tourists. Despite the fact that the 9/11 atrocity targeted specific buildings in New York City and Washington, DC, it created a widespread fear of flying in commercial aircraft all over the world for many months after September 2001.

The critically distinctive nature of this attack was that the hijacking of an airline morphed from a threat to passengers, crew and the aircraft to the employment of a fully fuelled aircraft as a weapon of mass destruction. As Figure 5.1 demonstrates, 9/11 caused a one-year dip in global aviation from September 2001 through to the end of 2002.

During the last quarter of 2001 and the first quarter of 2002, the number of people who flew internationally and domestically declined significantly as many people expressed a fear of flying due to the events of 9/11. The international aviation sector, dominated by its global governing body the International Civil Aviation Organisation (ICAO) and its main membership association International Air Transport Association (IATA), undertook a comprehensive review of global airline security, which also involved Airports Council International (ACI), the global association of the world's airports. The global airline security review was completed in late 2002 and headline measures were extensively publicised in the public media and widely disseminated to the tourism industry.

It is worth noting that terrorist attacks against airlines and airports continued to take place after 2001, the attacks on Brussels and Istanbul Airport in 2016 being two significant examples. However, after the aviation industry introduced stringent security measures for airlines and airports after 2001, the frequency and likelihood of such attacks diminished significantly.

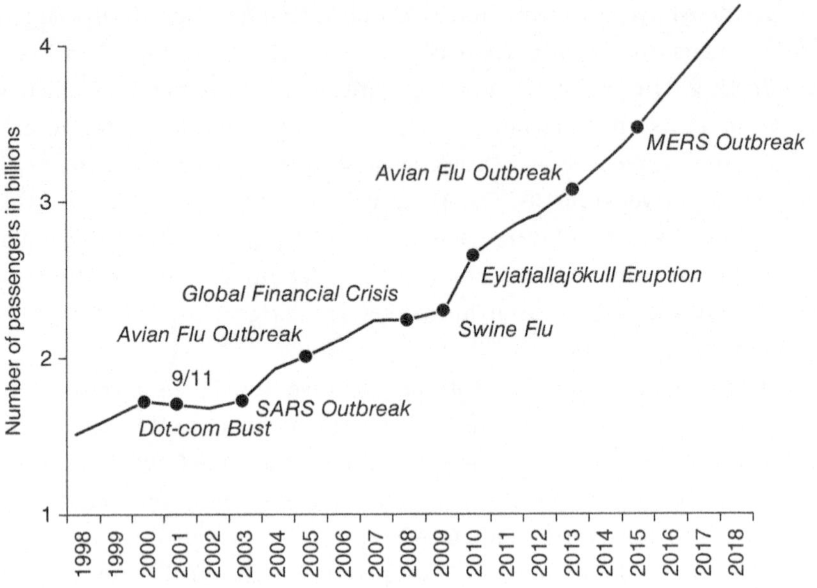

Figure 5.1 Global air passenger traffic, 1998–2018

Source: International Civil Aviation Organisation, 2020. Used with kind permission.

Although terrorist attacks against aviation declined after 9/11, those terrorists who targeted tourists found softer targets such as accommodation, attractions, resorts, markets, railway stations, shopping malls and public gathering places to name just some. The 14 July (Bastille Day) 2016 truck ramming attack in Nice, which killed 87 people and wounded 485, the Barcelona truck ramming attack in August 2017, which killed 13, and the London (Westminster Bridge) ramming attack of June 2017, represent three terrorist attacks targeting tourists in areas known by the perpetrators to be popular tourist visitation sites.

Although it may appear counter-intuitive, terrorist attacks don't necessarily lead to steep declines in tourism in the destination where an attack occurred. Terrorist attacks (even those targeting tourists), unless they are part of a series of attacks, occurring in Western countries rarely lead to a significant downturn in tourism. IS attacks in France during 2015 and 2016 slowed growth of international visitation to France but they did not result in a decline. In Western countries, security forces, the governments and the police are typically quick to reassure visitors and the media that security upgrades were implemented after an incident. In most cases, the perpetrators were rapidly captured or killed or as security personnel usually term it, 'neutralised' (Seabra et al., 2020).

However, terrorist attacks, especially those targeting tourists, can often have long-term negative perceptual impacts on tourism in developing countries. The Bali bombings of

2002 and 2005 had a significant impact on international tourism to Bali, which lasted for years (Gurtner, 2007). A series of terrorist attacks in which tourists were the primary target in Turkey during 2015 and 2016 led to a significant downturn in international tourist arrivals to Turkey (Karemalikli et al., 2020). Although tourism to Turkey ultimately recovered by the end of 2019, only to share in the global travails of Covid-19 in 2020, the Turkish government underwent a long and costly process to restore confidence about its safety as a tourist destination from its key source markets. A terrorist attack by gunmen operating on behalf of IS at the El Kantaoui coastal resort in Sousse (Tunisia) in 2015 killed over 30 British tourists out of a total of 38 people killed. The attack resulted in a virtual cessation of British tourism to Tunisia (BBC, 2015). Terrorist attacks targeting tourists to Egypt in 2014 and 2015 led to significant declines in tourism. In Egypt, the combination of political instability and terrorism since 2011 has hampered Egypt's ability to attract a recovery of international tourist arrivals to match the record level of 14 million arrivals in 2010.

Terrorism and the threat of terrorism figure prominently in government travel advisories. The response of governments and their security agencies in countries targeted by terrorist attacks (especially those incidents targeting tourists) often determines the severity and longevity of a negative travel advisory applying to a destination affected by terrorism. As travel advisories are treated as an extra-territorial security responsibility of travel generating countries to their citizens, foreign ministries treat the threat of terrorism in their travel advisories very seriously.

There is an increasing level of inter-governmental cooperation on assessing the terrorist threat to tourists in targeted destinations. This is clearly illustrated by the 'five eyes' intelligence cooperation between the governments of the UK, USA, Canada, Australia and New Zealand. The five 'Anglosphere' countries share intelligence on a range of military and security threats, including terrorism. The wording of government travel advisories – especially the security assessments, frequently reflect a high level of cooperation between the 'five eyes' governments (O'Neil, 2017).

In May 2017, the author was an invited speaker at the Asia Pacific Economic Cooperation (APEC) Counter-Terrorism and Tourism Working Group conference on terrorism and tourism held in Bali. APEC has a well-established counter-terrorism working group focusing on threats to tourism in APEC economies. APEC involves 21 governments encompassing the Pacific Rim. This includes the three 'superpowers', US, Russia and the People's Republic of China. It also includes Japan, Taiwan, Republic of Korea (South Korea), most South-East Asian nations, Canada, Australia, New Zealand, Papua New Guinea, Mexico, Peru and Chile. The commitment of this substantial group of governments responding to terrorism targeting tourism demonstrated a shared recognition that terrorism's threat to tourism impacts on a vital segment of the national economy in all of these countries.

Figure 5.2 Delegation leaders of the APEC Counter Terrorism Working Group Bali, 2017

Source: the author

Transnational approaches to tourism security with an emphasis on counter-terrorism are evident in the Association of Southeast Asian Nations (ASEAN) countries and the Caribbean countries, most of which have a high level of economic dependence on tourism (Avraham, 2015).

One of the barriers to establishing a global tourism industry response to terrorism involves the fragile political sensitivities of many countries in recognising the reality of terrorism. After the Bali bombing attack of 2002, the Indonesian government expended a great deal of diplomatic energy criticising Western countries (especially Australia) for issuing cautionary travel advisories to its citizens in reference to Indonesia. By contrast, the Indonesian government was initially resistant to enhancing tourism security measures.

Indonesia's hosting of the 2017 workshop and its active support for the APEC counter-terrorism working group demonstrates an important policy shift for Indonesia. The fact that representatives from Russia, China and the USA have a shared concern about the threat of terrorism to tourism is a positive sign. The working group supported a number of key action plans arising from the workshop and the deliberations with security heads from the 21 member countries. The APEC Counter Terrorism Working Group's strategic plan for 2018–2022 includes the following measures related to tourism (APEC Counter Terrorism Working Group, 2017):

- Work with the other APEC working groups (especially those related to trade) to build capacity and increase cooperation in the fields of secure trade and travel.

- Enhance the facilitation of technological measures to ensure tourism in the region is faster, easier and more secure.
- Enhance cooperation with the ICAO and the International Organization for Standardization (IOS) to improve the secure flow of travellers throughout the region.
- The Counter Terrorism Working Group will work with the Tourism Working Group to explore potential opportunities to improve protection for tourists, and share experiences and best practices in securing tourism destinations and travel, and products. A specific recommendation of the Bali 2017 workshop included pre-arrival security advice for travellers to all APEC destinations.
- Collaboration between the Counter Terrorism Working Group and the Maritime Experts Group working on maritime security to enhance port and maritime security.
- Enhance collaboration between the Counter Terrorism Working Group and the Aviation Experts Group as the lead working group on aviation security to enhance aviation and transport security.
- Counter Terrorism Working Group will work with other international organisations (including IATA, UNWTO, WTTC and IMO) to implement the Advance Passenger Information and Passenger Name Record (API and PNR) programmes to secure and facilitate travel within the APEC region.

Case Study 5.1

Egypt 1993–2011

In Egypt, the traditional Egyptian response to a series of terrorist attacks in the 1990s culminating in the 1997 Luxor massacre was to reassure tourists that Egypt would manage the security risk, without mentioning any specific measures.

However, the November 1997 massacre in which jihadist terrorists killed over 50 tourists was such a high-profile atrocity that the Egyptian government was obliged to visibly upgrade security for tourists, especially in the face of steep decline in tourism arrivals (26 per cent) in 1998. Prior to the massacre, the Egyptian Tourism Authority (ETA) contracted the services of US-based consultancy firm Burson-Marsteller to develop a post-terrorism marketing strategy. It was only effectively employed after the massacre (Beirman, 2003).

Martin Langford, who was placed in charge of working with the ETA by Burson–Marsteller in 1993, established a strategic approach to post-terrorism tourism recovery, which has been widely accepted as a template for many countries seeking to restore tourism following a terrorist incident (Pizam and Mansfeld, 1996).

Langford's key management principles for post-terrorism tourism recovery were:

1. Define the real problem, short- and long-term.
2. The media, however hostile, is rarely the problem; rather it's best to focus on tangible measures to restore confidence. These would have included upgraded security protocols.

(Continued)

3. Centralise and control the information flow by appointing one spokesperson. Avoid lies and distortions, respect media deadlines, avoid speculation and control interviews and press briefings.
4. Isolate a crisis management team from extraneous business concerns.
5. Contain the problem. In the case of terrorism, define the actual extent of the problem.
6. Depend on no one individual fully and ensure the management team is accountable to the tourism ministry or government head office.
7. Resist the urge to adopt a combative stand in media briefings and understand that the media is looking for the cause of the terrorist attack, or a party to blame for the crisis. Understanding this involves spokespeople resisting the temptation of inadvertently providing leads in this direction.
8. Identify allies and stakeholders with whom you can work and will support your case.
9. Clarify what specific measures are undertaken in the destination to address the problem without revealing some specific tactics that may warn potential perpetrators.

Government tourism ministries and tourism associations have added many strategies and measures to their recovery toolkit since the 1990s. However, many aspects of Langford's approach are included in counter-terrorism strategies of national tourism boards today. The advent of social media sites in computers and hand-held devices, coupled with widespread personal access to the worldwide web, has transformed media since the 1990s.

Today, any individual with a smart phone can disseminate a media message. The idea of a destination or corporate spokesperson, exercising full control of communicating crisis recovery, is largely redundant. Currently, most enlightened destination management organisations (DMOs) incorporate a multi-platform traditional and social media strategy as a core element in communicating with stakeholders to advance reputation management and destination recovery. The key recovery challenge for any destination affected by terrorism is to clearly explain and illustrate the distinction between negative perceptions and facts. While this applies to recovery from all forms of crises, managing perception has become especially important in responding to terrorism. There is extensive research evidence that prior to the advent of Covid-19, terrorism aroused a higher level of fear for prospective tourists than virtually any other form of threat to tourism.

The commitment of government and national tourism industry leadership to acknowledge and recognise the threat of terrorism in countries is an important determinant of recovery.

Case Study 5.2

Egypt 2011–2020

Until the anti-government uprising of 2011, Egypt was the most visited country in the Arab world. In 2010 a record 14.2 million international tourists visited Egypt, spending

US$12.5 billion. Tourism was the leading source of Egypt's foreign exchange earnings. From the Luxor massacre of 1997 until 2011, Egypt experienced few outbreaks of terrorism directed at tourists. The exception was Sinai where some resorts, especially those popular with Israeli visitors, including Taba and Dahab, were targeted by Islamist groups.

Although Egypt adopted a planned approach in response to terrorism in the 1990s as discussed above, there was little evidence that this approach was emulated during and after the outbreak of uprisings in Egypt in 2011. The so-called 'Arab Spring' uprisings centred on Cairo to oust the regime of Hosni Mubarak in February 2011 succeeded, largely due to the support of Egypt's military for their immediate aim of ousting Mubarak who assumed the Egyptian leadership after the assassination of Anwar Sadat in October 1981. In the immediate aftermath of Mubarak's resignation, there was a sense accompanied by a hope on the part of younger, educated Egyptians that a new democratic Egypt would emerge.

The demonstrations in Cairo and the ultimate ousting of Mubarak during February 2011 were far more a response to an episode of political instability than terrorism. Its immediate impact on tourism was a mass exodus of foreign tourists aided by over 30 governments who charted planes to facilitate the rapid evacuation of their citizens from Egypt, and worked closely with Egyptian and foreign tour operators to achieve this. The timing of the protests occurred during the height of Egypt's peak tourism season.

The ETA painted the Mubarak overthrow as a flowering of democracy and their international marketers urged tourists to return and spend their money to contribute to the flowering of a new democratic Egypt. There was no accompanying guarantee of tourist safety and the marketing campaign was more aspirational than realistic.

During 2011, tourism to Egypt plummeted in both visitor numbers and expenditure (9.4 million visitors representing a 30 per cent decline on 2010 levels and an expenditure of US$8.7 billion, or just over 30 per cent decline).

The hope for a democratic Egypt foundered by default. The political vacuum created by Mubarak's overthrow was unfilled by an organised and united political movement, committed to liberal or democratic values. In Egypt there has never been a significant organised political movement committed to liberal democratic values. By contrast, the conservative Islamist Muslim Brotherhood (despite being outlawed by successive Egyptian regimes since Gamal Abdel Nasser came to power in 1954) had an extensive underground political and clerical network of Islamic supporters in Egypt's main urban centres, and the towns and villages throughout regional and rural Egypt.

Following the overthrow of Hosni Mubarak in February 2011, the military ruled Egypt until a national election. Government restrictions against the Muslim Brotherhood were suspended and the Muslim Brotherhood, led by Mohamed Morsi, organised supporters for the election all over Egypt. The students and intelligentsia who actually drove the protests were politically idealistic but lacked a national political organisation and had little support outside Egypt's two largest cities, Cairo and Alexandria.

When elections were finally held in October 2012, the Muslim Brotherhood won convincingly. The Morsi government lasted for two years before it was overthrown by the

(Continued)

military and Abdel el-Sisi was installed as president. During the Muslim Brotherhood's brief period in power, there were numerous attacks on Egypt's minority Coptic Christian community and other non-Muslim groups. Even after the military resumed power, Egypt's internal political instability was accentuated by acts of jihadist terrorism in Sinai. Some of these attacks fatally targeted tourists. The most severe of the terrorist attacks targeting tourists in Egypt killed 224 Russian tourists when a bomb planted by jihadists in a Russian commercial aircraft exploded in mid-air after the plane took off from Sharm El Sheikh in October 2015. These events combined to shape negative perceptions of Egypt as an undesirable tourist destination for some years. On November 26, 2017, an attack on a mosque in Northern Sinai killed over 300 Egyptian worshippers. While tourists were not the target of this attack, atrocities on such a large scale reinforced the perception of Egypt as a dangerous destination for tourists. The ETA's failure to substantively address or refute negative perceptions about the safety of Egypt since 2011 resulted in a patchy recovery of tourism to Egypt since 2011, which has been gradual, uneven and prolonged.

TERRORISM, RACISM, THE PROLIFERATION OF WEAPONS AND THE THREAT TO TOURISM

Terrorism is more than an existential threat to tourists. Many of the concepts at the heart of tourism's *raison d'être* (open minds, open hearts and respect for diversity) are the antithesis of the core beliefs of most terrorist groups. In recent years, multiple terrorism attacks carried out by jihadist groups targeted tourists. These include the truck ramming attack in Nice on the Bastille Day holiday in July 2016, which killed 87 people and wounded 485 (mainly tourists). Between 2015 and 2019, jihadist attacks targeted tourists in Tunisia (2015), Paris (2015), Burkina Faso (2016), Brussels (2016), Istanbul (2016), Orlando, Florida (2016), Sharm El Sheikh, Egypt (2016), Berlin (2016), Westminster Bridge, London (2017), Manchester (2017), London Bridge (2017), Nairobi (2019) and Colombo, Sri Lanka (2019). These represent just a small portion of the many Islamist jihadist attacks targeting tourists in recent years. In the context of tourism risk, it is notable that terrorism is a widespread and global threat to the viability of tourism in many countries (Seabra et al., 2020).

However, terrorist attacks that either target or impact on tourism are not confined to Islamist groups. One of the bloodiest recent terrorist attacks in Europe was perpetrated by a self-styled white supremacist. Anders Breivik killed 77 people in two attacks in Norway in July 2011. Most of his victims were young socialists attending a vacation camp on an island near the capital, Oslo. In March 2019, a terrorist attack by Australian white supremacist Brenton Tarrant, targeting Muslim worshippers in two mosques in Christchurch, New Zealand, killed 51 people and wounded 40 in the worst terrorist attack in New Zealand's history. In the USA, white supremacist groups over a period of 120 years have targeted African Americans, Jews, Muslims, Gay people and Latinos. While tourists are not the specific targets of these groups, the targeting of minority groups generates an atmosphere of fear to visit destinations affected by such attacks, especially among the targeted racial or ethnic groups. In 2018, a shooting attack on a Pittsburgh Jewish synagogue by a self-avowed white supremacist led to the death of 11 worshippers and the wounding of seven. In 2019, an armed white supremacist attacked a Walmart store in El Paso, killing 23 people and wounding 23. Most of the victims and the declared targets were Latino Americans.

Racial and religious bigotry is at the heart of many terrorist attacks. The attitude of racial or religious hatred runs counter to the ideals of tourism's engagement with destination communities as expressed by the International Institute for Peace Tourism (IIPT, 2020), which promotes respectful engagement between tourists as guests and destination communities as hosts.

A key enabler for mass terrorist and criminal attacks (especially in the USA) is the ready availability of guns, especially illegal assault weapons. Although many terrorist attacks involve weapons other than guns, including truck or car ramming, bombs and bladed weapons, guns tend to play a dominant role in terrorist attacks against tourists. In most countries, guns, and especially semi-automatic and automatic weapons, are difficult to obtain legally. Consequently, most terrorists operating in Western countries tend to acquire their weapons via illegal channels.

The USA is in an unusual position by contrast with most countries in the world. As the powerful US gun lobby group, the National Rifle Association (NRA) (2020), constantly reminds US opinion leaders, the second amendment of the USA's constitution includes the right of citizens to bear arms. There is constant debate in US political and media circles on whether that 'right' includes a justification to own weapons that could literally massacre a football crowd in a matter of seconds. Although character checks do apply to gun licences, there are differences between the 50 states of the USA in terms of their stringency. Gun related crime and terrorism in the US has included mass-murders in schools, places of worships, shopping malls and city centres. Attempts by successive US administrations, including Barack Obama's administrations (2008–2016), to enact more restrictive gun laws and tighter gun licensing provisions are routinely met with acrimonious rejection from the NRA and other gun lobby stakeholders. The relative ease of obtaining assault weapons in the USA while arguably not a cause of terrorism, is clearly an enabler. During the 2020 US presidential election campaign, there were concerns about the presence of armed militias from both the right wing and left-wing extremes. This raised the possibility that bullets would be exchanged rather than angry words in American politics, adding to the verbal toxicity of political discourse in the USA between supporters and opponents of President Donald Trump's administration between 2017 and 2021. From a tourism perspective, there was concern that tourists could be inadvertent victims of exchanges of politically motivated fire.

THE SECURITY CHALLENGES INVOLVED IN PREVENTING AND RESPONDING TO TERRORISM TARGETING TOURISM

Terrorists have a number of significant tactical advantages when mounting an attack against a chosen target. The terrorist can choose the time, place, target and mode of attack. Security professionals or the military can establish defensive strategies, personnel and technology, which may deter a terrorist attack or even respond to an attack, but it would be unrealistic to suggest that terrorist attacks are preventable. Tourism enterprises and tourism infrastructure, including airports, cruise ship, bus and train terminals, hotels, attractions, shopping malls and markets, are all potential terrorism targets and often require varied security approaches.

After the 9/11 attack in September 2001, the global airline industry and airports joined forces under the auspices of the International Civil Aviation Organisation (ICAO), International Air Transport Association (IATA) and Airports Council International (ACI) to develop an integrated aviation security strategy to minimise the likelihood of a repeat of 9/11. During 2002, the author attended two of the many meetings that aviation leaders held around the world to discuss this vital issue for the security of global aviation. As both aviation and airports are subject to strict global standards, a comprehensive security strategy was developed. Some measures agreed were obvious necessities lacking on most airlines prior to 9/11. The most important was keeping the cockpit door locked during a flight. The ease in which terrorists could enter an unlocked cockpit was a common security threat during 9/11 and in many preceding hijackings. There was a broad range of measures agreed, including enhanced screening of passengers and their baggage prior to moving to airside at airports. Airports upgraded their covert and overt security measures on both landside and airside. For reference: landside is that part of the airport open to the public, which can be used for meeting, greeting and socialising before the departure of passengers and after arrival; airside is that part of the airport that's open only to passengers screened by security and registered airline, airport and government staff (including immigration, customs, police and military).

Many of the security measures agreed by the aviation industry were technological in nature and all arrangements were subject to review. The significant outcome was that while terrorist organisations were aware that the aviation sector had upgraded security there, they were not aware of the specific measures taken and most attempted terrorist attacks on commercial aviation after 9/11 were thwarted until the 2015 terrorist attack against a Russian aircraft in Sharm El Sheikh, Egypt. In this case, the bomb that was secreted onto that aircraft was the work of a licensed airport employee who later identified himself as a supporter of IS.

A broad challenge for all sectors of tourism centres on the broader tourism security dimension. When airlines and airports became a more challenging target for terrorists after 9/11, terrorists sought to attack what they considered to be softer targets. A favoured target choice for terrorists targeting tourists are international hotels and resorts in major gateway cities. Major hotels attracted an international clientele, many wealthy and influential local people, and as centres of hospitality, the security measures tended to be discrete but porous.

Figure 5.3 illustrates the key security challenges faced by tourists and the tourism industry. Each element of tourism has its own set of security problems. The issue of tourism policing is discussed in some detail in Chapter 8, as many measures to detect, prevent and respond to terrorism are similar to those used to deal with crime. Apart from hotels, tour groups in coaches, open attractions (such as markets, plazas and shopping malls), places of worship, historic sites, railway and coach terminuses have become favoured targets for terrorists in recent years. In the latter instances, terrorist attacks are as likely to kill or injure tourists as they are local people. However, individually or in combination, they create a sense of fear and panic among prospective tourists (Beirman, 2018).

A full discussion on the range of security measures to avert or respond to terrorism would fill a very thick book. In essence, security measures fall under two broad categories. The first is overt security. This includes bollards and barriers, visible screening procedures, uniformed and visible security personnel and visible surveillance equipment. Some countries employ tourist police whose role is more as a help resource for tourists rather than a protective role, but it is argued that their presence is a deterrent to crime and terrorism.

The second broad category of security measures are the covert measures. Concealed screening, ID checks and surveillance, plain clothed security personnel, alarms and a range of high-tech security measures, including locks, concealed barriers and traps and bullet-proof glass windows, are just some of the measures employed. In 1968, when the Israeli airline El Al became one of the first airlines in the world to carry plain clothed, armed sky-marshals onboard flights, it became a game among passengers to spot the sky marshal.

The key to effective counter-terrorism security requires a high level of collaboration between tourism, hospitality and events businesses, emergency management agencies and the appropriate tier of government (Beirman, 2011). The need for tourism security that terrorism has generated has led to the establishment of a number of specialist agencies that have expertise on tourism security (Tourism and More, 2020; Ridley 2020).

The most vulnerable targets for terrorists to strike are those links in the tourism security chain where public access is unrestricted or subject to minimal overt control. Shopping malls, open markets and publicly accessible tourism attractions are prime targets for mass terrorist attacks.

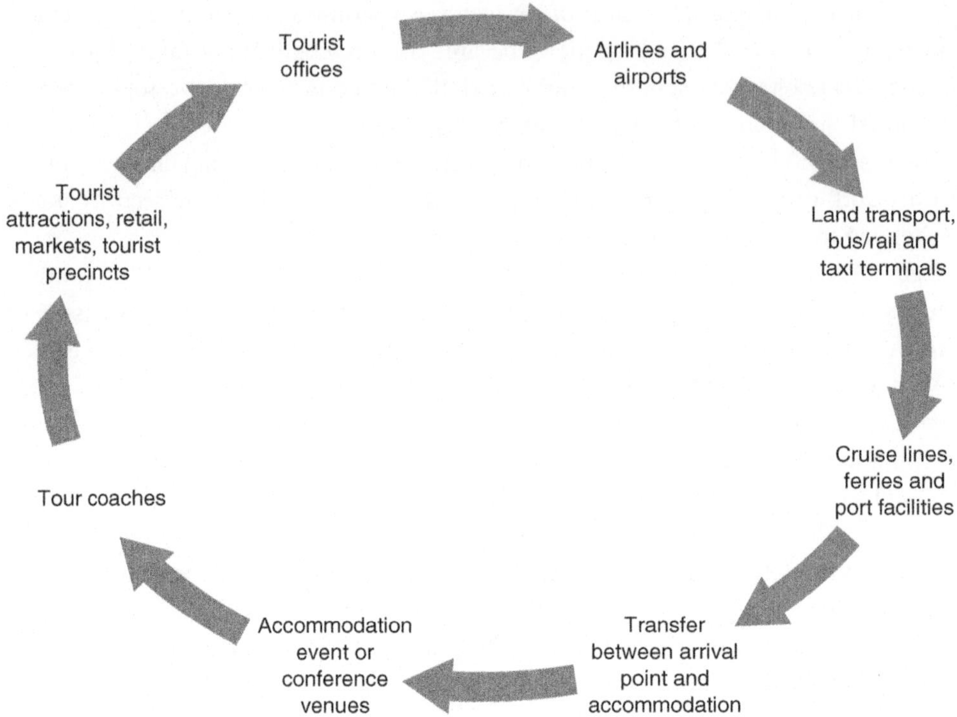

Figure 5.3 The tourism security cycle

Source: Beirman, 2018

Although it is technically possible to seal hotels from public access, this rarely occurs unless there is an imminent threat of a terrorist attack. The accommodation sector is well aware that a high proportion of mass casualty terrorist attacks in recent years have targeted luxury central business district (CBD) hotels. From a terrorist's perspective, five-star CBD hotels are a target rich environment with many people from a variety of countries. Causing death and mayhem in such a hotel is an almost certain guarantee of global headlines, which is exactly what most terrorists seek. Large, central city hotels face a significant dilemma from a security perspective.

In addition to catering to the security of paying 'in-house guests', most major hotels host meetings and events ranging from boardroom meetings to conferences and weddings attended largely by people who are not in-house guests. In addition to guest rooms, these hotels have restaurants, health clubs, retail shops and other facilities. For many people, including locals and tourists, the hotel lobby is an easily identifiable meeting place. Hoteliers and their customers tend to regard hotels as centres of hospitality and there is an understandable reluctance on the part of most hotel managers to engage in the scale of intrusive screening and security, readily acceptable at airports.

Many hotel chains, including Hilton, Marriott, Accor and the Intercontinental Hotel Group, have invested heavily in technological security and surveillance measures to monitor threats and to remotely screen guests and their baggage. Certain events held in the hotels, especially those involving VIPs, can be literally sealed off from the remainder of the hotel. Most hotels employ plain clothed security staff and seek to ensure that their security presence is as discreet as possible. Many leading hotels have certain premium floors set aside for guests who can only access those floors, their assigned rooms and shared facilities such as lounges.

The 2008 attack on the Taj Mahal Palace Hotel in Mumbai (India), in which terrorists from the jihadist Lashkar-e-Taiba group attacked and killed guests and staff at the hotel as part of a multi-target terrorist attack in Mumbai, India, raised a range of counter-terrorism issues for all hotels in the world. These included the question of whether hotel security staff should be armed, the nature of co-operation between police, military and hotel security, and whether the entrances to certain hotels (likely to be targeted by terrorists) should be subject to screening.

A disturbing recent trend has been for terrorists to book as paying guests at a hotel to smuggle weapons, including suicide bomb vests, assemble them in rooms and detonate them in the hotel. This tactic was first employed by jihadist terrorists at the JW Marriott Hotel in Jakarta in 2009 and most recently by Islamist terrorists in three Colombo (Sri Lanka) hotels in April 2019. These incidents have increased the pressure on hotels to more carefully screen guests' baggage.

Case Study 5.3

The April 2019 Terrorist Attacks in Colombo, Sri Lanka

Tourism to Sri Lanka has been obliged to overcome a range of crises over 40 years. The most prolonged crisis for modern Sri Lanka was the long-running civil war between the Sinhalese majority and the Tamil minority between 1983 and 2009. Sri Lanka gained independence from Great Britain in 1948, and from that time until 1972, was known as the Dominion of Ceylon. In 1972, a nationalist government declared the country as the Republic of Sri Lanka.

There are many ethnic and religious groups in Sri Lanka. The dominant group is the 70 per cent of the population who are Sinhalese, most of whom are Buddhist by faith. The next largest group are the Tamils (most of whom originate from Southern India and are predominantly Hindu). Sri Lanka is also home to substantial Muslim

(Continued)

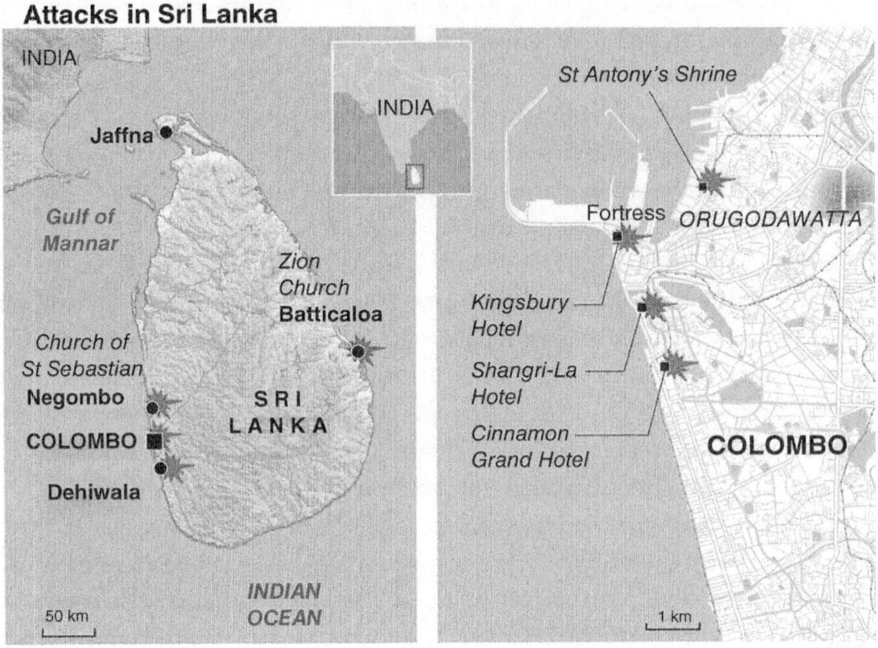

Figure 5.4 Terrorist attacks in Sri Lanka, 21 April 2019

Source: Agence France Press, 2019. © Agence France Presse/ Laurence Saubadu, Vincent Lefai.

and Christian communities. While the Sinhalese dominate in about 75 per cent of the island country, Tamils are a minority concentrated in the north-east tip of Sri Lanka (Szczepanski, 2019).

Ethnic tensions have long simmered between the Sinhalese and the Tamils. In the years following the declaration of the Republic of Sri Lanka, the government sought to impose Sinhalese language and culture on the entire population. This in turn spurred a secessionist movement among the Tamils. Between 1983 and 2009, there were episodes of bloody conflict between the government and the Liberation Tigers of Tamil Eelam (LTTE), more typically referred to as the Tamil Tigers.

The conflict included terrorist attacks, largely conducted by the LTTE, which targeted tourists and tourism infrastructure and were intended to dissuade tourists from visiting Sri Lanka. In 1995 and 1996, major hotels were targeted in the capital Colombo. On 24 July 2001, a Tamil Tigers' attack on Colombo International Airport destroyed most of an Air Sri Lanka fleet of passenger aircraft, killing many people (Beirman, 2003).

In addition to the ongoing civil war, Sri Lanka was among the most severely affected countries from the Indian Ocean tsunami of late December 2004. Over 35,000 people were killed in Sri Lanka, a small portion of them tourists travelling along the country's coast or staying at waterfront hotels and resorts. Second only to Ache Province in Indonesia, Sri Lanka suffered the most severe damage and the highest rate of casualties resulting from the tsunami.

Between 1995 and 2009, inbound international visitation to Sri Lanka struggled to reach 500,000 people per annum. The combination of ongoing civil war and occasional terrorist attacks targeting tourists deterred most intending travellers from including Sri Lanka in their travel itineraries. The Sri Lankan Tourism Board (2020) energetically marketed the destination globally, but it appeared that whenever they were gaining some headway, another reputation sapping terrorist incident would occur. By 2008–2009, the Sri Lankan government's military forces gained the upper hand in the fighting against the LTTE at the cost of over 100,000 combatants and civilians killed on both sides of the conflict (including over 47,500 Tamil civilians). For a country with a population of 22 million, the casualty rate on both sides was severe (Szczepanski, 2019).

In 2009, the two sides finally came to a political settlement. While residual resentment would remain for years to come, fighting between the Tamils and the central government ceased. From the cessation of the civil war in 2009, tourism to Sri Lanka blossomed. Travellers from all over the world were enthusiastic to view an exotic but hitherto troubled destination. By 2018 (see Figure 5.5) over 2.3 million international tourists visited Sri Lanka, an all-time record and five times the number who visited in 2009.

Confidence in the future growth of Sri Lankan tourism was shattered by a multi-pronged jihadist terrorist attack using a mixture of guns, grenades and suicide bombers, which targeted three major Christian churches, the Church of St Sebastian, St Antony's Shrine (both near Colombo) and Zion Church in Batticaloa on Sri Lanka's east coast. The attacks took place on 21 April 2019, Easter Sunday (one of the holiest days on Christianity's calendar). In addition to the attacks on churches, three luxury hotels in the capital, Colombo, Kingsbury Hotel, Shangri-La Hotel and Cinnamon Grand Hotel were struck. The hotel attacks involved suicide bombers who were paying guests at the three hotels and who assembled bombs in their rooms. The attacks on the churches killed 220 people and who the attacks on the hotels killed 39 people. Altogether, the highly coordinated attacks killed 259 people and injured over 500. IS claimed responsibility for the attack. Part of the claim, and arguably a fanciful part, was that this attack was a revenge attack for the Christchurch (New Zealand) terrorist attack on worshippers in two mosques during the previous month.

As discussed, the lack of coordination between national counter-terrorism intelligence, the military and the police in Sri Lanka were contributing factors to the

(Continued)

commission of the 21 April terrorist attacks and the high number of casualties. Although 39 of the 259 people killed in the attacks were tourists, these attacks (focused as they were on leading hotels) clearly damaged the reputation of Sri Lanka as a tourism destination. Tourism is one of Sri Lanka's most important and lucrative economic sectors. Until 21 April 2019, tourism was the third largest component of Sri Lanka's national economy and employed 1.5 million people. As the figures below attest, in the six months since the Sri Lanka attack, tourism arrivals to Sri Lanka between May and December of 2019 were significantly lower than the numbers in the corresponding months of 2018.

NB	2018	2019	Change
January	238,924	244,239	+2.2%
February	235,618	252,033	+7.0%
March	233,382	244,328	+4.7%
April	180,429 NB terrorist attack occurred on 21 April 2019 166,975		–7.5%
May	129,466	37,802	–70.0%
June	146,828	63,072	–57.0%
July	217,829	115,701	–46.9%
August	200,359	143,587	–28.3%
September	149,087	108,579	–27.2%
October	153,128	118,743	–22.5%
November	195,582	176,743	–9.5%
December	253,169	241,169	–4.5%
Jan-Dec	2,333,796	1,913,796	–18.0%

Figure 5.5 International tourism arrivals in Sri Lanka, corresponding months January–December 2018 and 2019

Source: Sri Lanka Tourism Development Authority, 2019; Sri Lanka Ministry of Tourism, 2020. Figure by the author

The above figures represent a significant decline of just over 400,000 international tourism arrivals since 21 April 2019. Although the negative month on month figures eased as the time distance from the terrorist attack increased, it is clear that 2019 was a year of decline in tourism, which will have a negative impact on tourism employment and a ripple effect throughout the Sri Lankan economy.

In the eight months since the April 2019 terrorist attacks, the decline in tourism revenue to Sri Lanka between May 2019 and October 2019 (compared with the same period in 2018) was conservatively estimated as US$850 million. The estimated direct job losses in Sri Lanka's tourism industry between April 2019 and December 2019 were 41,000.

The figures for loss of revenue were based on an average tourism per capita expenditure to Sri Lanka of US$1,840 and job loss is based on one job per US$30,000 in revenue (UNWTO, 2014). This is a very conservative estimate of the loss of direct tourism jobs and does not take into account the indirect loss of jobs in economic segments closely linked to tourism.

A significant and prolonged decline in tourism increased unemployment and resumed the cycle of Sri Lankans seeking illegal channels of emigration to neighbouring countries, including Australia. One of the great challenges that the Sri Lankan government and tourism industry leadership have faced centres on the rebuilding of trust among all stakeholders that the country is a safe tourism destination. Visible and well publicised actions to enhance security and restore destination reputation following significant acts of terrorism, especially those targeting tourists, are essential preconditions for rebuilding the confidence and trust of international visitors. Consumer confidence in visiting tourism destinations is always undermined when a major security lapse occurs.

Generally, developing countries (including Sri Lanka) tend to experience a long and slow recovery (1–2 years) after major terrorist attacks targeting tourists. This has been the experience for Turkey (2015–2017), Tunisia (2015), India (2008), Bali (2002 and 2005), and Egypt (1997 and 2011–2019). Tourism in developed countries such as the UK, Spain, Belgium and France has also been affected by terrorist acts. However, tourism recovery in developed countries tends to be far quicker, as security services in developed countries are perceived to be more agile and responsive in response to terrorist incidents than countries in the developing world.

Sri Lankan tourism was certainly showing signs of recovering by the end of 2019 but in common with all countries, it too was impacted by the even greater challenge of Covid-19, which affected Sri Lanka from March 2020.

The response of Sri Lanka to the terrorist attack involved two key dimensions. The first was a government instigated root and branch review of counter-terrorism security in the country. The second was a coordinated marketing campaign run by Sri Lankan tourism and involving all sectors of Sri Lanka's tourism industry. Extensive use was made of social media with significant emphasis on targeting media and tourism industry professionals from Sri Lanka's key international source markets.

Tourism Sri Lanka emphasised that security in hotels and all sectors of the Sri Lankan tourism industry were being upgraded. Travel agents, travel journalists from all media platforms and tour operators were invited to undertake familiarisation tours to Sri Lanka to see for themselves and report back to their respective stakeholders. Sri Lanka ensured it had a very prominent presence at the PATA Tourism Mart held in Astana (now Nur-Sultan), Kazakhstan in September 2019, and the World Travel Mart held in London in November 2019.

SUMMARY AND CONCLUSION

Terrorism has proven to be a significant threat to tourism during the 21st century. Although the threat posed by terrorism to tourism was overshadowed in 2020 by the Covid-19 pandemic, terrorism remains a significant threat to tourism destinations and businesses alike.

Despite the advances in counter-terrorism security applying to all sectors of the tourism industry, tourists and tourism remain vulnerable to terrorist attacks. Publicity is the oxygen of most terrorist groups irrespective of their political and social cause, religion or ideology. Striking at physical or human targets who can guarantee headlines and media coverage in multiple countries is a core reason why terrorists have a distinct preference for targeting international tourists in hotels, planes, nightclubs, coaches, airports and events.

This chapter has not discussed event security at length, but it's noteworthy that the largest single expense for organisers at mega events around the world is security, most of which is directed at protecting the event, its participants and attendees from the threat of terrorism. Cruise operators especially those that operate in maritime choke points are highly conscious, having teams to engage in counter-terrorism. It is a very short step from piracy for material gain to maritime terrorism threatening a cruise ship with thousands of passenger and crew.

Tourism security is increasingly a collaborative venture between tourism businesses and organisations and the various arms of emergency management, police and at times the military. Terrorists are always developing new tactics and counter-terrorism faces a constant challenge to anticipate existing and future potential threats. As this chapter has demonstrated, there is significant growth in the level of international cooperation on counter-terrorism matters as terrorism is a threat to all states in the world. As the APEC example shows, China, Russia and the USA have many political and diplomatic differences but they do effectively cooperate in the field of counter-terrorism in relation to tourism.

Although terrorist attacks are damaging to the reputation of destinations and tourism enterprises, an effective response to terrorism can ultimately present an opportunity to restore confidence in the destination and business, and to take measures that minimise the likelihood of future attacks. Recovery of tourism destinations from terrorism needs to stress a contextual element. A terrorist attack on a specific target in a specific locality should not lead to an assumption often made by the more sensationalistic sections of the media that an entire national destination is unsafe. An important part of crisis communications, especially in relation to terrorism, is the establishment of context. The response of global airlines and airports to the 9/11 attack of 2001 has enhanced the safety of aviation. No tourism destination or business can promise that terrorism can be eliminated as a threat, but enhancing both proactive and reactive risk management measures reduces the likelihood of terrorism.

Discussion Questions and Issues

1. Hotels are clearly a major target for terrorist attacks. Discuss the various security measures that a hotel can take which would minimise its impact on the dignity and privacy of guests while affording all hotel users a high degree of security in the hotel.
2. Based on the case studies on Egypt in this chapter, what should the Egyptian tourism industry be doing to create a sense of assurance for prospective travellers to Egypt?
3. High speed rail is one form of international transport that is taking root in Europe and Eastern Asia. Based on what you have read in this chapter, discuss a range of security measures that high speed rail operators can take to reduce exposure to terrorism.
4. Engagement with stakeholders is a critical element in restoring the reputation and consumer confidence in a destination after a major terrorist incident. Based on an example of your choice, map out a strategic reputation recovery plan for a destination that has experienced a major terrorist attack.

REFERENCES

Asia-Pacific Economic Cooperation (APEC) (2017) *Strengthening Tourism Business Resilience Against the Impact of Terrorist Attack*. Report of the APEC Counter-Terrorism Working Group's Workshop on 9–10 May 2017, Bali, Indonesia. APEC Counter-Terrorism Working Group, APEC Secretariat: Singapore.

Avraham, E. (2015) 'Destination image repair during crisis. Attracting tourism during the Arab spring uprisings', *Tourism Management*, 47: 224–32.

BBC (2015) *Terrorist Attack on Tourists in Tunisia*. https://bbc.com/news/world-africa-33287978. Accessed 23 December 2020.

BBC (2019) *Sri Lanka Attacks: What We Know of the Easter Bombings*. https://bbc.com/news/world-asia-48010697.

Beirman, D. (2003) *Restoring Tourism Destinations in Crisis. A Strategic Marketing Approach*. Allen & Unwin: Crows Nest (Sydney), Australia.

Beirman, D. (2011) 'The integration of emergency management and Tourism', *The Australian Journal of Emergency Management*, (Australian Government Attorney General's Department). 26(3): 30–34.

Beirman, D. (2018) 'Tourism crisis and safety management', in Cooper, C., Volo, S., Gartner, W.C. and Scott, N. (eds), *The Sage Handbook of Tourism Management*. Sage: London, UK. pp. 154–68.

Gurtner, Y. (2007) 'Crisis in Bali: Lesson in tourism recovery', in Laws, E., Prideaux, B. and Chon, K. (eds), *Crisis management in tourism*, Cabi Publishing: Wallingford, Oxon, UK. pp. 81–97.

International Air Transport Association (IATA) (2020) www.iata.org. Accessed 7 December 2020.

International Civil Aviation Organisation (ICAO) (2020) www.icao.int. Accessed 7 December 2020.

International Institute of Peace Tourism IIPT (2020). https://peacetourism.org. Accessed May 15 2021.

Karemalikli, H., Kahn, A.A. and Karimi, M.F. (2020) 'Is terrorism a real threat to tourism? Analysis of inbound and domestic tourism in Turkey', *Current Issues in Tourism*, 23(17): 2165–81.

National Rifle Association (NRA) (2020) https://home.nra.org. Accessed 17 December 2020.

O'Neil, A. (2017) 'Australia in the five eyes network: the perils of an asymmetric alliance', *Australian Journal of International Affairs*. https://doi.org/10.1080/10357718.2017.1342763. Accessed 30 November 2020.

Pizam, A. and Mansfeld, Y. (eds) (1996) *Tourism, Crime and International Security Issues*. John Wiley & Sons: New York. pp. 120–50.

Ridley, A. (2020) *Security and Risk Management Science*. https://linkedin.com/in/tonyridley?locale=ja_JP. Tony Ridley is one of the world's leading private consultants in the field of Tourism Security Management. Accessed 7 December 2020.

Seabra, C., Reis, P. and Abrantes, J.L. (2020) 'The influence of terrorism in tourism arrivals. A longitudinal approach in a Mediterranean country', *Annals of Tourism Research*, 102811.

Sri Lanka Tourism Board (2020) https://srilanka.travel. This site contains a great deal about Sri Lankan tourism's response to risk and crisis. Accessed 16 December 2020.

Sri Lankan Tourism Statistics (2020) https://sltda.gov.lk/statistics. Accessed 7 December 2020.

Szczepanski, K. (2019) *The Sri Lankan Civil War*. https://thoughtco.com/the-sri-lankan-civil-war-195086. Accessed 20 December 2020.

Tarlow, P. (2014) *Tourism Security: Strategies for Effectively Managing Travel Risk and Safety*. Elsevier: Amsterdam. pp. 98–100.

Tourism and More (2020) Tourism Security Consultancy. www.tourismandmore.com. Accessed 7 December 2020.

UN World Tourism Organisation (2014) *Asia Travel Trends*. UNWTO Madrid and Global Tourism Economy Research Centre. https://www.e-unwto.org/doi/pdf/10.18111/9789284416301. Accessed 8 May 2021.

6

THEME 3: DESTINATION RECOVERY FROM NATURAL DISASTERS

INTRODUCTION

The pleasure periphery (Turner and Ash, 1975; Weaver and Lawton, 2014) refers to those areas of the world in which sun, sea and sand are the dominant attractions especially for leisure tourists. This encompasses destinations as varied as the Greek islands, the South of France, the Italian Riviera, Spain, Florida and the Gulf of Mexico region, Florida and Hawaii in the USA, the Caribbean island nations, Pacific islands, much of coastal South-East Asia, northern Australia, Sub-Saharan Africa, Mexico and South and Central America. Collectively, these represent most of the popular tourism destinations on Earth. Concurrently, they also represent destinations that are exceedingly vulnerable to natural disasters.

Natural beauty and a warm tropical climate can be major assets for the desirability and marketability of tourism destinations, but a location in the pleasure periphery is accompanied by significant risks and dangers. Many of these erstwhile idyllic destinations can suddenly be transformed into danger zones, exposing tourists and tourism infrastructure to extensive damage from tropical storms (cyclones, typhoons and hurricanes). In some parts of the world, extreme climatic events are regarded as a risk and hazard rather than a potential crisis because they are predictable and frequent, annual, seasonal occurrences. Monsoonal storm events are common in many tropical maritime regions.

However, natural disasters in the pleasure periphery extend well beyond the impact of cyclonic storms. Many of the world's most desirable tourism destinations are also located on or near tectonic fault lines (the associated Pacific regions being referred to as the 'Ring of Fire'), areas where earthquakes and volcanic eruptions are most likely to occur.

Some of the most destructive natural disasters in history have resulted from earthquakes, volcanoes and their marine by-product, tsunamis. Some of these events have levelled cities and extensively destroyed large regions. Ironically, some volcanoes are actually tourist attractions (usually when extinct or dormant), but even active volcanoes have attracted tourists. Unfortunately, even those active volcanoes that are deemed to be simmering, and therefore safe to visit briefly, can erupt unexpectedly. New Zealand's Whakaari (White Island volcano) was a case in point. On 9 December 2019, a group of cruise tourists visited White Island on an excursion as part of a cruise on the Royal Caribbean Cruise ship *Ovation on the Seas*. While visiting the volcanic island, members of the excursion party were trapped by an eruption that cut off avenues for escape, killing 21 people (Benton, 2019).

In the case of all natural disasters, the scale, timing, extent and impact of the disaster is random. Although volcanologists, meteorologists and seismologists can analyse and forecast the activity of volcanoes, earthquakes and the weather, even their best forecasts have a significant temporal, spatial and intensity margin of error.

Natural disasters are not limited to the warm climes of the pleasure periphery. Tourists in alpine regions face the risk of avalanches, glacial contractions, mudslides and extremely cold and biting winds, which are endemic to these regions. Of all the many risks that impact on tourists and tourism, natural disasters, whether they arise from drought, flood, weather events or Earth's movements, contain the shared elements of unpredictability and severity. Even though many tourism destinations and businesses prepare for natural disasters, the scale of some natural disasters exceeds the capacity of the most effective human planning.

This chapter focuses on the recovery of tourism destinations and associated enterprises affected by natural disasters. The prime focus is on natural disasters arising from extreme climatic events, fire, flood, tectonic movement, volcanoes and land or snow slippage. Some disasters, frequently defined as natural disasters, actually have a significant human element associated with both causality and spread. Bushfires and forest fires represent a case in point. Although there have been many fire events triggered by natural causes such as lightning strikes, extreme heat and atmospheric dryness, many of the largest bush/forest fires include a significant human element. These range from deliberate acts of arson at worst, to electrical faults, 'controlled back-burning' that goes out of control and careless acts, such as failure to douse a barbeque or careless disposing of a lit cigarette butt. Bushfires are discussed in this chapter. However, while bush/forest fires represent a significant crisis, they are part of that grey area between natural and human caused disasters.

RISKS AND CRISES WHICH ELICIT SYMPATHY AND SOLIDARITY

Unlike human created crises, such as political instability, crime and terrorism, destinations and destination communities affected by natural disasters tend to elicit a great deal of sympathy from a portion of tourists who are motivated to show solidarity with affected destinations, especially if they have a familial, ethnic, emotional, aspirational or business connection with that destination. The phenomenon of solidarity tourism is more likely to be evident in response to natural disasters than any other form of crisis. The very randomness of a natural disaster and the knowledge that every place on Earth is potentially vulnerable to acts of nature (or 'acts of God' to use insurance parlance) tends to arouse a sense of empathy.

Today, there are many online resources to maintain a brief on impending and active natural disasters at any point in the world. Among the best and most comprehensive is the Global Disaster Alert and Coordination System (GDACS) (2020). The UN World Tourism Organization (UNWTO) was among the first global tourism associations to introduce a natural disaster tracking system when it launched the Tourism Emergency Response Network (TERN) in 2006 (UNWTO, 2020).

Case Study 6.1

The 2004–2005 Indian Ocean Earthquake and Tsunami – Focus on Thailand

On 24 December 2004, a massive undersea earthquake just off the coast of Ache Province in Northern Sumatra (Indonesia), measuring 9.1 on the Richter scale, triggered a tsunami that generated waves which inundated coastal areas of the Indian Ocean, from the epicentre in Ache Indonesia to Somalia on the North-East African coast, some 7,000 kilometres to the west. The Indian Ocean tsunami is the most destructive natural disaster of the 21st century (to date) and caused over 230,000 deaths, mostly in Ache Province, Indonesia. However, in Thailand, Sri Lanka, India and even in Kenya, which was 6,500 km from the earthquake's epicentre, tourists were among the casualties. Almost half of the 5,350 people killed in Thailand as a consequence of the Indian Ocean tsunami were international tourists. For many Northern Europeans escaping a cold winter, Phuket in southern Thailand is a popular sun and sea tourist destination. Most of the tourist casualties in Thailand's Andaman Sea coasts and islands were from

(Continued)

Northern Europe with over 500 deaths recorded among tourists from each of Sweden and Germany (Gurtner, 2007).

The death of a significant number of tourists from many countries magnified media coverage of the tsunami, effectively transforming it into a global disaster. Although many more people were casualties of the Indian Ocean tsunami in India, Sri Lanka and Indonesia than was the case in Thailand, the high proportion of tourist deaths in Thailand generated a far higher level of global media coverage than any other affected area.

Since the outbreak of the Asian Financial Crisis in 1997–1998, Thailand has proven to be one of the world's most resilient tourism destinations. It has successfully maintained growth in tourism arrivals under the most challenging circumstances.

Figure 6.1 Indian Ocean tsunami

Source: © Agence France Presse/ROSLAN RAHMAN

The Indian Ocean tsunami of 26 December 2004 was one of the greatest challenges for Thailand from a tourism recovery perspective. In Thailand, the number of deaths attributed to the tsunami was far lower than Indonesia, Sri Lanka and India but Thailand experienced the highest proportion of tourist deaths of any country affected by that catastrophic event. Just over 40 per cent (or 2,464) of the 5,395 people who were confirmed dead from the tsunami in Thailand were international tourists who had come from over 40 countries to enjoy a tropical Christmas period. The 500 dead from Sweden and Germany topped the international tsunami fatalites in Thailand (Beirman, 2018). The tsunami sites in Thailand were concentrated on its relatively short Andaman Sea coast. However, this coast includes some of Thailand's most popular tourism attractions notably Phuket and Phan Nga Province (north of Phuket) and the Phi Phi Islands (south-east of Phuket).

From the day of tsunami, the Tourism Authority of Thailand (TAT) worked closely with the Pacific Asia Travel Association (PATA) and the UNWTO to develop and implement a tourism recovery strategy and programme. The large number of casualties and the extensive destruction and damage to homes, businesses and accommodation meant that the top priority for local and national government agencies focused on a triaged approach to victims and a prioritised approach to the assessment of damage to property, transport sanitation and power infrastructure.

PATA had the advantage of its global headquarters being located in Bangkok, relatively close to the headquarters of the TAT. In the days immediately following the earthquake tsunami, TAT and PATA quickly provided information on the status of those parts of Thailand affected by the tsunami. This included details of casualties and damage, and equally importantly, the details of operable tourism and transport infrastructure. The author supplied PATA with a simple checklist for affected tourism businesses on recovery measures, which could be taken to restore businesses, their reputation, stakeholder communication and their marketability. PATA released, free of charge to all its members and tourism businesses in Thailand, copies of its 2003 tourism risk and crisis management manual (Van Walbeek and The Winning Edge, 2003). The PATA website extended its coverage of tourism status updates to all tsunami affected countries including Myanmar, Sri Lanka, Indonesia and India.

TAT was careful to avoid promotional marketing of affected areas while rescue, relief and victim identification and assistance were occurring. However, it did actively promote tourism to the large parts of Thailand, unaffected by the tsunami. One of the main perceptual problems for countries affected by a natural disaster is that perceptions created by extensive media coverage suggest that an entire country is a disaster zone. While Ache and many parts of southern Sumatra incurred extensive damage from the tsunami, most of Indonesia, including Indonesia's most popular destination of Bali, was unaffected. Bali did experience a brief drop in tourism after the tsunami. In crisis management parlance, the isolation strategy (identifying the problem area in relation to a wider context of most of the destination as unaffected) was employed by destination marketing bodies and industry stakeholders in both Thailand and Indonesia.

One month after the Indian Ocean tsunami occurred in January 2020, the UNWTO convened an 'emergency meeting' of all its member states in Phuket to discuss a broader strategy of global tourism assistance for all countries affected by the Indian Ocean tsunami. Representatives from 70 countries attended.

The UNWTO developed an overall strategic approach to support those destinations affected by the tsunami. The strategy included an extensive marketing campaign and fundraising to assist affected tourism businesses. In essence, this was very similar to the strategy developed by PATA nearly a month before but with the added benefit of a global tourism industry commitment at government level. Of the affected countries, only the government of India refused to accept assistance from the UNWTO.

(Continued)

India chose to use its own resources for relief and recovery and while welcoming the offer of UNWTO assistance, decided to go it alone. However, Thailand, Indonesia, Sri Lanka and Myanmar stated they were prepared to accept support. Thailand's President Thaksin Shinawatra visited the UNWTO meeting to personally thank all delegates for their countries' support of Thailand. Part of that support involved the travel industry media pledging to extensively cover and promote destination recovery and marketing of the affected destinations. The delegates (including the author, at that time appointed to represent Israel's Ministry of Tourism) were taken on an inspection tour of Phuket organised by TAT. They witnessed that even in Phuket, some towns and beach resort areas suffered extensive damage from the tsunami but many others were relatively untouched. Such is the random nature of a natural disaster.

While Thailand did experience a small downturn in tourism during 2005, the overall strategy of isolating the problem from the broader context tended to work well. Some months after the tsunami when much of the damage to Phuket was cleaned up and most infrastructure repaired, tourism in Phuket was actively promoted from the second half of 2005. Foreign aid and support from the Thai government and the global tourism industry helped many tourism-related businesses ride out the financial damage of the disaster.

NATURAL DISASTERS AND TOURISM

The impact of natural disasters on international tourism and tourists generally magnifies the extent of online and traditional media attention. Floods, fires, drought, mudslides, volcanoes, earthquakes, storms and tsunamis cross the line from being a localised disaster to a tourism crisis when tourists are affected by them or are among the casualties. Nepal's 2015 earthquake killed over 9,000 people, including 89 international tourists (UN Dispatch, 2015). While tourists represented just under one per cent of total casualties, the impact of this natural disaster on tourists from multiple countries generated a high level of global media coverage.

In late 2019, the impact of natural disasters on tourism continued to make headline news. The 2018 and 2019 earthquakes and aftershocks in Lombok (Indonesia), a major earthquake and tsunami near Pala in Sulawesi, heatwaves in Europe, forest fires in California, floods in Kerala (India), earthquakes and typhoons and volcanic eruptions in Japan are just some of the current natural disasters that influence tourism flows to affected regions. In the case of Lombok, a popular Indonesian island destination for backpacker tourists just east of Bali, a severe earthquake on 5 August 2018, followed by multiple aftershocks, triggered a mass evacuation of tourists from the island (BBC, 2018).

Although the 2018 Lombok earthquake resulted in 563 deaths (virtually all local residents) and widespread destruction of homes and infrastructure, a reading of the Australian media placed far greater emphasis on the impact of the earthquake on Australian tourists than on local citizens. One of the ongoing issues in addressing tourism recovery from sites of natural disasters is that tourists and the media in source markets tend to be far more concerned about the welfare and safety of their own citizens than the welfare and safety of local residents. Despite the global nature of media coverage, media outlets appealing to a specific audience can be exceedingly parochial to the point of insensitivity to the problems of people outside their audience reach.

In early November 2020, a massive typhoon struck the Philippines, a country which frequently experiences highly destructive typhoons. The most damaging typhoon to hit the Philippines since the 19th century was the category five, Super Typhoon Haiyan (or Yolanda), which struck the northern Philippines in November 2013. Over 6,300 people died, mainly Filipinos; an estimated 25 tourists died as a direct consequence of Typhoon Haiyan. The Philippines Ministry of Tourism undertook a major post-crisis recovery programme, which focused on making a clear distinction between those parts of the Philippines' 7,000 plus islands affected by the typhoon and those which were not. This is part of a fundamental 'isolation strategy' employed by DMOs during natural disasters to contextualise the extent of a disaster's impact (Khazai et al., 2018; Beirman, 2003) on a destination region.

Some segments of the global tourism market respond positively to destinations that have experienced natural disasters. Increasingly, we are seeing the phenomenon of 'solidarity tourism' in which some tourists want to assist the recovery of communities, which have incurred damage resulting from a natural disaster. The rapid recovery of Nepal's tourism following the April 2015 Gorkha earthquake and subsequent aftershocks has, to a large extent, been driven by growth of volunteer tourism (Beirman et al., 2018; Wearing et al., 2020) This was dominated by many altruistic young visitors who chose to devote all or part of their visit to Nepal to assist communities rebuild and recover from the damage caused by the quake. The concept of solidarity tourism also extends to expatriate nationals of destinations that experienced a natural disaster. Many people who have family, community and national ties to a destination that had experienced a natural disaster are drawn to visit and volunteer their services as an expression of their solidarity with the destination community.

At a global and transnational level, the tourism industry has worked energetically to provide moral and marketing support to destinations that have experienced natural disasters. In 2012, the World Travel and Tourism Council (WTTC), the UNWTO and PATA joined forces to stage the Global Tourism Summit of April 2012 in Japan. The summit was a mark of support for Japan's tourism industry, a year after the destructive earthquake tsunami of March 2011, which resulted in the deaths of close to 20,000 people (UNWTO, 2020).

Symbolically, the first day of the summit was held in Sendai, the Japanese prefecture that bore the brunt of the tsunami's destruction. In 2017, PATA and the South Pacific Tourism Organisation held their first regional summit in Vanuatu to symbolise support to Vanuatu's recovery from Cyclone Pam in 2015. In fact, the UNWTO, WTTC, PATA, Association of Southeast Asian Nations (ASEAN) Tourism, the APEC Tourism Forum, the European Tourism Commission and Caribbean Tourism Organization have frequently expressed or provided support to destinations affected by natural disasters. This is not merely an act of altruistic benevolence. The support of destinations that have experienced natural disasters involves mutual self-interest based on a realisation that many natural disasters, especially storm-related disasters, affect multiple countries. As already mentioned in this chapter, nature does not respect human political borders and many national disasters evolve into transnational crises. The fact that the wrath of many natural disasters is random in temporal, spatial and social terms means that any community anywhere in the world is vulnerable to the caprice of nature's fury. The concept that natural disasters involve shared potential risks tends to arouse a greater level of external sympathy for destinations affected by natural disasters than other causes of destination crises.

Responses to some major natural disasters in the world indicate that citizens of many countries are willing to assist the affected countries with medical aid, reconstruction and rescue assistance. In those instances, when foreign citizens (most of whom are tourists) are casualties of the natural disaster, the likelihood of this aid being made available is enhanced.

Case Study 6.2

Volunteer Tourism and the Recovery of Nepal after the 2015 Earthquake

On 25 April 2015, a massive earthquake struck central Nepal and caused extensive damage to 15 per cent of the country, including its most populous area centred on the capital city Kathmandu. Over 9,000 people (including 89 foreign nationals) died as a direct consequence of the quake and the aftershock of 14 May.

After the Nepal earthquake and aftershocks of April–May 2015 the majority of international rescue, medical and construction teams, which came to Nepal to assist in rescue, medical treatment and reconstruction, came to Nepal from countries whose nationals were among the casualties of the earthquake. While all these teams devoted most of their efforts to assist affected Nepalese communities and citizens, there was significant focus placed on the welfare of expatriate victims.

Volunteer tourism played a significant role in Nepal's post-2015 earthquake recovery. In the months immediately following the main quake of 25 April 2015, the call for volunteers emphasised a need for people with practical, trade and professional skills who could be put to work for rebuilding, medical treatment and infrastructure redevelopment. From 2016, more generic forms of volunteer support were invited and welcomed. As Wearing et al. (2020) and Upadhayaya (2018) demonstrated, volunteer tourism played a significant role in attracting tourism to Nepal after the earthquake to the extent that by 2018, international tourism arrivals to Nepal had achieved all-time record levels. The recovery of tourism to Nepal elicited support from selected international celebrities and VIPs. These included Britain's Prince Harry (the Duke of Sussex) who visited Nepal in 2016 and called on the youth of the world to demonstrate their commitment to Nepal's recovery by visiting as volunteer tourists. Prince Harry was widely reported assisting a rebuilding programme in an earthquake-affected village during his visit. Nepal had an unusually longstanding and vibrant volunteer tourism infrastructure, which enabled short and long-term placements for volunteer tourists.

Although volunteer tourists are a small portion of total tourist visitation to Nepal, their high level of commitment and relatively lengthy stays motivated many of them to act as unofficial tourism ambassadors for Nepal. They actively encouraged many young and adventurous tourists from all over the world to visit Nepal in statements and video clips hosted by Nepalese tourism social media sites. There is no doubt that Nepal's rapid recovery was partially facilitated by volunteer tourists relating their positive experiences of Nepal on social media. Although international visitor numbers to Nepal declined in 2015 (the year of the quake) by 2017, as indicated in Figure 6.2, international visitor numbers had exceeded all previous records. During 2018, international tourism numbers to Nepal exceeded one million for the first time ever (Government of Nepal, 2019).

Unlike Nepal, many countries, including Indonesia, which lacked a comparable volunteer tourism sector, were unable to leverage the degree of volunteer tourism support for recovery programmes after natural disasters.

In May 2015, PATA's head office, in close cooperation with Nepal's Ministry of Tourism and the member companies and staff of PATA's Nepalese Chapter, established PATA's Nepal Rapid Recovery Taskforce. The role of the taskforce was the development of an extensive marketing strategy to advance the tourism recovery of Nepal via a social media driven recovery marketing campaign. The taskforce report, published in June 2015 (PATA, 2015) was the basis for Nepal's rapid tourism recovery in the two years following the quake. A key reason for the high level of acceptance of the PATA taskforce was the fact that it was a collaborative effort between PATA's international team of tourism crisis and recovery specialists and leaders of the Nepalese tourism industry. At all stages, the Nepalese team members had the final say on each strategic decision made. A key lesson that all team members (including the author) learned was that when residents and citizens of a destination have ownership of a recovery programme, it's more likely to be embraced and adopted. The core messages of the

(Continued)

recovery campaign during 2015–2016, which are common to recovery campaigns following many natural disasters, were:

1. Nepal invites you! Tourists are warmly welcome.
2. The destination is largely safe to visit apart from specified locations.
3. Most of Nepal (85 per cent of the country) was undamaged by the earthquake.
4. The airport and most hotels are open.
5. If you care about Nepal and want to make a real contribution to Nepal's recovery, come and visit, or visit and volunteer. [A call for solidarity tourism.]
6. Visitors coming to Nepal are having a great experience. See and hear what they have to say.
7. [Celebrity visitor/volunteer testimonials from Susan Sarandon (USA), Jackie Chan (Hong Kong) and Prince Harry (UK) (Beirman et al., 2018).]

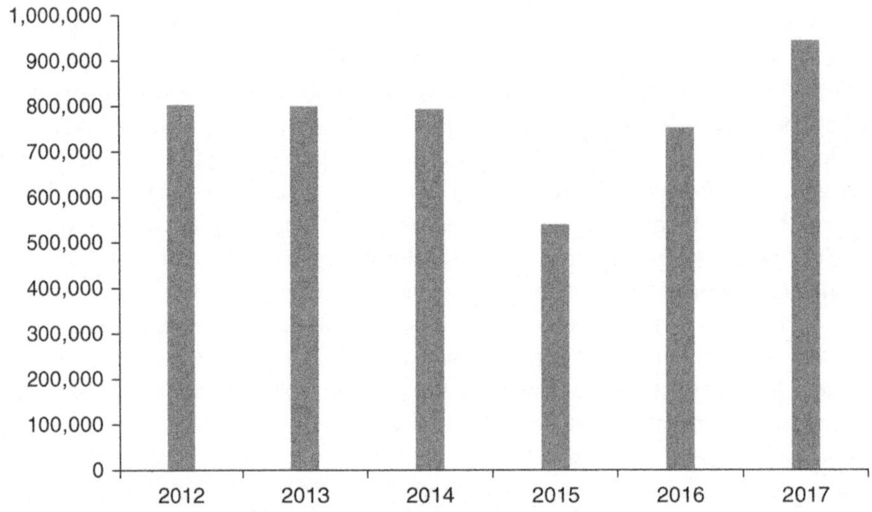

Figure 6.2 International tourist arrivals, Nepal 2012–2017

Source: Government of Nepal Tourism Statistics (MoCTCA), 2018. Graph developed by the author

PREPARING FOR NATURAL DISASTERS: JUST BUILDING BACK, OR BUILDING BACK BETTER AFTER NATURAL DISASTERS?

No country on Earth is immune to national disasters. In 2020, there were floods in Europe. Typhoons, hurricanes and cyclones threatened coastal communities in many parts of the world. In Greece and Turkey, earthquakes caused destruction in the Turkish resort city of Izmir and the Greek island of Samos. Western USA experienced massive forest fires while the eastern half of the USA experienced floods. Many natural disasters are annually recurring, some are completely unexpected. Most have an impact on tourists and tourism.

Proactive risk management preparedness for natural disasters is vital in terms of national planning and tourism infrastructure. Analysing climate, seismology and vulcanology data can assist in predicting the likelihood of a natural disaster occurring. Even with a range of technologies, including earth-orbiting satellites, meteorology, seismology and vulcanology are not exact sciences, and at best can provide a guide to the likelihood of a natural disaster but cannot predict with certainty.

At the Best Education Network (a global network of tourism academics) meeting held in Kingston, Jamaica, in June 2005, 40 tourism academics bore witness to a remarkable presentation by a Jamaican meteorologist, Mr Franklin McDonald (2005). Mr McDonald analysed the shifts in hurricane paths over the past 200 years and correctly concluded that Jamaica, which in the past had been spared exposure to hurricane paths in the Caribbean Sea, was by 2005 a prime hurricane target. He also demonstrated that the data indicating landing points on the southern USA's Gulf of Mexico coast had shifted west and predicted that New Orleans was a highly likely target. One month later, Hurricane Katrina struck New Orleans, inflicting immense damage and casualties on a city, much of which is below sea level and protected by an antiquated system of dykes (designed to withstand a force 3 storm), which was grossly inadequate to protect the city from a force 5 hurricane. It is almost certain that US meteorologists had access to the same data, which begs the question, frequently asked in the USA after Hurricane Katrina, why weren't US emergency management officials ready for a Katrina event? Risk management for destructive mass-casualty natural disasters involves the employment of massive resources, the expenditure of vast amounts of money, and the political will to spend the money and utilise the resources. The USA is one of many countries globally that tends to react to crisis events rather than prepare effectively for them (Blakely, 2011).

By contrast, Japan is among the most risk-ready countries on Earth. Geographically and geologically, Japan is vulnerable to a wide range of natural disasters, including earthquakes, volcanos, tsunamis, floods, typhoons and landslides, which occur frequently in the mountainous and heavily populated island nation. Japan has invested heavily in natural disaster defence and preparation. In Japan's tourism industry, all hotels have evacuation procedure guides in multiple languages, assembly points and crisis management drills for everything from fire, flood and tsunamis, to volcanic ash clouds and earthquakes. Following the Kobe earthquake of 1995, building regulations throughout Japan were upgraded to ensure that all public and private homes were built to strict earthquake resistant standards. Despite all these measures, much of the Sendai Prefecture of North-East Honshu was inundated by a massive tsunami in March 2011, triggered by a massive earthquake measuring 9 on the Richter scale. The quake and the tsunami exceeded the scale of any previous earthquake in Japan and overwhelmed the defence mechanisms developed by the Japanese. Almost 20,000 people died as a result of the tsunami and the Fukushima nuclear power plant located on the sea shore experienced a near meltdown, which restricts its operation to this day. Keeping this in mind, when the Okinawa Prefecture developed its risk management plan during 2012, they operated on the basis of the historically worst-case disaster (Takamatsu, 2014).

Lessons learned from past natural disasters serve as a guide for post-crisis recovery. Although it may seem like a cliché, the term 'building back better' is a key guide to post-disaster recovery. Sandy Hollway, who was in charge of organising the 2000 Sydney Olympics, was also appointed by the Australian government to head two destination recovery programme, one in 2003 to coordinate Australia's capital city Canberra's recovery from massive bushfires and the second in 2006 to assist far North Queensland to recover from Cyclone Larry. Hollway was a powerful advocate for building back better. In responding to both disasters, he proposed a range of enhancements to building regulations, local government zoning, vegetation management, drainage, firebreaks and a re-marketing programme to minimise the repetition of damage arising from a similar disaster and focusing on rebuilding the safety and reputation of the affected destinations (Prideaux et al., 2008).

A similar principle applied to post-tsunami tourism resort planning in Thailand after the 2004 Indian Ocean tsunami, which included tsunami warning systems, siting of coastal accommodation to the highest practical ground level and in multi-storey accommodation, ensuring that the ground floor level was not utilised for guest accommodation, and evacuation routes (Gurtner, 2007).

Not all destinations adopt a national policy of building back better. On 29 September 2009 at 6.48am local time, a major earthquake measuring 8.1 on the Richter scale generated a massive tsunami that devastated the southern coast of the Samoan islands of Upolu and Savaii and parts of American Samoa, including the capital city Pago Pago. On the southern coast of Samoa's most populous island, Upolu, home to a number of Samoa's most popular beach resorts, many of these resorts sustained severe damage and casualties, including tourists. Over 190 people were killed due the tsunami. There were 149 deaths in Samoa, 34 in American Samoa and nine in Tonga. Additionally, over 3,000 Samoans were rendered homeless, a devastating loss for a country with 182,000 people. In some Samoan resorts, the most luxurious suites were located at the water's edge and in some cases over the water. At Sinalei, an upmarket resort located on Upolu's southern coast, the presidential suite and the honeymoon suite were destroyed by the tsunami and its occupants were among the casualties.

Samoa has an office of the UN Disaster Management Centre in the capital Apia, which has tracked cyclones, hurricanes, earthquakes and volcanic activity for many years. The centre received a warning from the Pacific Tsunami Warning Centre in advance of the 2009 tsunami. However, the notice was received only 15 minutes in advance of the tsunami striking the coast and there were few measures in place to deal with a rapid evacuation from the area affected by the tsunami. Evacuation, procedures, alarms, sirens and shelter arrangements on elevated ground were in place in the capital, Apia. But beyond Apia and the international airport, there was very little government-developed emergency management infrastructure. However, the management of a number of south Upolu Samoan resorts, on their own initiative (including Sinalei and Virgin Cove), established evacuation paths and assembly points in preparation for cyclones, hurricane and tsunamis. Apia, located on the north coast of Upolu, the opposite coast from the tsunami event, was unaffected by the tsunami (Pararas-Carayannis, 2009).

Although Samoans experienced considerable suffering and distress as a result of the tsunami, the tsunami response of most Samoan citizens was in accordance with national emergency drills. By contrast, many tourists from affected resorts and accommodation providers simply panicked and there was little in the way of official advice to assist them to seek safety or evacuate. A notable exception was Apia's famous and venerable Aggie Grey's Hotel, which literally opened its doors to assist stranded and confused tourists and some locals during and after the tsunami.

Much of this was due to a remarkable woman who set the standard for hotel-based emergency management in Samoa. Beverly Barlow held the title of Reservations Manager at Aggie Grey's Hotel from 1997 until her death in 2012. Although it was outside of her official brief, Mrs Barlow, a former Samoan College Teacher (and Head of English), organised a set of detailed risk and crisis management strategies for the hotel and its staff. She personally trained and drilled all staff members to deal with a range of risks, which could affect the operation of the hotel and its guests. She maintained a watch on natural disasters, independent from Samoa's National Disaster Management Centre. She also maintained relations with Samoa's emergency management agencies and personnel. Mrs Barlow and the Aggie Grey's hotel staff assisted hundreds of dazed, confused and distressed tourists affected by the tsunami. The hotel offered temporary sanctuary and fed and watered many tourists seeking to evacuate Samoa. Mrs Barlow even organised buses to transport tourists from affected resorts on the south coast of Upolu to Aggie Grey's hotel and then, in cooperation with the airlines, transport to the airport. Effectively, Mrs Barlow and her colleagues provided the type of assistance that government emergency management agencies in Samoa were unable or under-equipped to provide. Mrs Barlow was an outstanding example of a tourism professional with an innate appreciation of the role of risk management in hotels. In February 2012, following floods that resulted from the breach of a dam wall during Cyclone Evan, Aggie Grey's staff rescued dozens of guests when parts of the hotel were inundated by flood water. The emergency management training regime for hotel staff developed by Mrs Barlow was followed by the hotel's staff and saved many lives.

The Samoan government's lack of a strategic approach to tourism risk and recovery manifested itself in the post-crisis rebuilding of tourism infrastructure after the tsunami. Seaside hotels and resorts were rebuilt after the tsunami but in exactly the same sites and in the same way as they were before the tsunami, leaving them vulnerable to a similar disaster whether it came in the form of a tsunami, cyclone, sea surge or hurricane. Samoa is far from the only country in the South-West Pacific that continues to build and site coastal resorts in response to market demand as opposed to risk minimisation.

In February 2010, Chris Flynn, the then Pacific Director of PATA, organised a conference held in Samoa under the title of Tourism Investment for the Development of Enterprise and Sustainability. The conference, sponsored by the European Union's Pro€ Invest, was attended by 150 delegates (including most of Samoa's tourism industry leaders) and intended to rebuild confidence in 'destination Samoa', and discuss strategies to enhance tourism risk and crisis management in Samoa and other island nations in South-West Pacific. Although many useful strategies were discussed by a line-up of outstanding speakers, including Chris Flynn, very few of these ideas were subsequently acted upon by the Samoan government and the tourism industry, even though the conference was held just a few months after the tsunami (Flynn C, pers. comm. 25 Oct 2017). Apathy towards translating risk management best practice into action is a significant problem in many countries.

Leadership and political will are core requirements when a nation or its tourism industry is committed to build back better after natural disasters. As some of the examples discussed above demonstrate, commitment and leadership at both government and tourism industry levels are vital to ensuring that tourism destinations and businesses actually treat risk and crisis as an opportunity to build back better after a natural disaster. The spirit of Beverly Barlow should inspire political and tourism industry leaders to take the initiative to lead when officials fail.

Case 6.3

Bush and Forest Fires – Australian Cases, 2018–2020

Tathra, New South Wales, 2018

The concept of solidarity tourism and communal resilience plays an important role in destination recovery from natural disasters. Bush and forest fires are not always natural disasters. While many can start from natural causes such as lightning strikes or extreme heat, there are many that are not natural in origin. Unfortunately, many bush and forest fires start from deliberate acts of arson. However, the progress and extent of a fire usually exceeds the intention of the people who may have deliberately ignited a fire or carelessly tossed a lit cigarette butt into the bush or forest. During 2018–2019, the author and a University of Technology Sydney (UTS) colleague, Associate Professor Stephen Wearing, worked with the Chamber of Commerce from the town of Tathra on the south coast of New South Wales in researching the town's tourism recovery from a destructive bushfire, which occurred in March 2018 (Beirman et al., 2019).

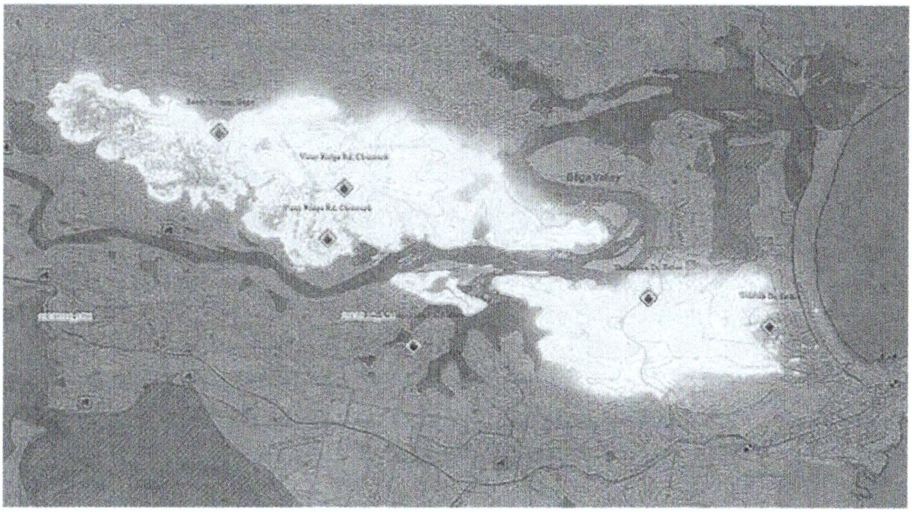

Figure 6.3 Extent of the March 2018 Tathra bushfire on the township and surrounding national parks (light shaded areas were fire affected)

Source: Image courtesy of NSW Rural Fire Service, 2019. Used with permission.

Tathra is a small coastal town, very popular with Australian family travellers who are attracted to its beaches, cycle ways and bushwalks. Many of Tathra's visitors stay during the school holiday periods at the caravan parks, serviced apartments and other accommodation, and the family groups tend to be repeat visitors. On 18 March 2018, severe bushfires swept through Tathra and the surrounding national parks, destroying 70 residential homes, damaging another 35, burning large areas of bushland and causing extensive damage to the town's largest caravan park. Fortunately, there were no deaths or injuries as a result of the fire, but the loss of a home and personal treasures is devastating to all who experience it.

(Continued)

Extensive and graphic news reports in the Australian and international media of the fires, burning homes and the people affected by it spurred an outpouring of public sympathy for the Tathra community.

The Tathra and District Business Chamber of Commerce (representing most of the town's businesses), in conjunction with local government (Bega Valley Shire Council) and the regional tourism bureau, Sapphire Coast Tourism, worked collaboratively to maintain tourism. In a highly effective campaign, Tathra's core family holiday market was encouraged to maintain their support for Tathra by visiting. Tathra's local community was supportive of those residents who had lost property, and businesses obtained a great deal of support from the local council to expedite insurance claims to cover losses that resulted from the fire. One of the key messages that Sapphire Coast Tourism communicated to all stakeholders was that the people of Tathra welcomed visitors and maintaining visitation was the most effective way of expressing support. In order to attract tourists while Tathra was going through its recovery phase, several new events were added to the annual range of town events.

Tathra's prime concern was maintaining continuous engagement with the local community to ensure that they were supportive of welcoming tourists, and maintaining a positive approach to the town's recovery. In various meetings, the residents of Tathra met with speakers from other parts of Australia who had experienced bushfires in their communities to understand that every bushfire experience is different and local residents respond differently to them.

Tathra has been successful in recovering from the 2018 bushfire. Despite the damage to the town, many of the natural attractions in the surrounding national parks, Tathra's sense of optimism and the support of its core tourist markets have enabled it to be a model of destination resilience. In September 2019, the Tathra Sapphire Coast Tourism Resilience Project, led by the Tathra and District Business Chamber, won the NSW Resilient Australia Award (Business Category) in recognition of the town's commitment to resilience. The project report, largely written by Suzette Fullerton and Mark Darby (2019), is an excellent example of a whole of community approach to tourism resilience following a natural disaster.

The tourism recovery model developed by Tathra in 2018 served as a template for many Australian regional and rural towns (some within 50km of Tathra), which experienced the threat from a far larger bushfire outbreak during late 2019 and early 2020.

Australia's Black Summer, 2019–2020

Australia is frequently afflicted with bushfires. The massive bushfires of late 2019–early 2020 (the most extensive in Australia's recorded history) raged in five Australian states. The fires destroyed millions of hectares of forest. In New South Wales, 60 per cent of the state's forests were affected. The scale of the fires led to smoky skies in Australia's two largest cities, Sydney and Melbourne, and the capital city Canberra. Preparing for and responding to bushfires has become a core risk management issue for regional Australian tourism and emergency management agencies. In 2012,

Dr Stuart Toplis from Tourism Victoria led the development of Tourism Victoria's bushfire preparedness and response strategy, which has served as the template for all other local and state tourism boards in Australia (Victorian Government Tourism Resources, 2020). The state of Victoria was the setting for Australia's most deadly bushfire in March 2009 (the Black Saturday Bushfire), which killed 173 people, injured over 400 and destroyed over 2,000 homes.

The sheer scale, speed and ferocity of Australia's 2019–2020 bushfire outbreak made it very difficult to save property, especially when fire fighters were operating on multiple fronts in specific regions. The cutting of many main roads by the fire posed a special challenge for both fire fighters and civilians seeking access and evacuation. The highest priority was the saving of human life, and considering the extent of Australia's fires indicated in Figure 6.4, the direct death toll from the fires of 34 people was remarkably low. However, it is estimated that about 440 died from smoke inhalation as an indirect consequence of the fires.

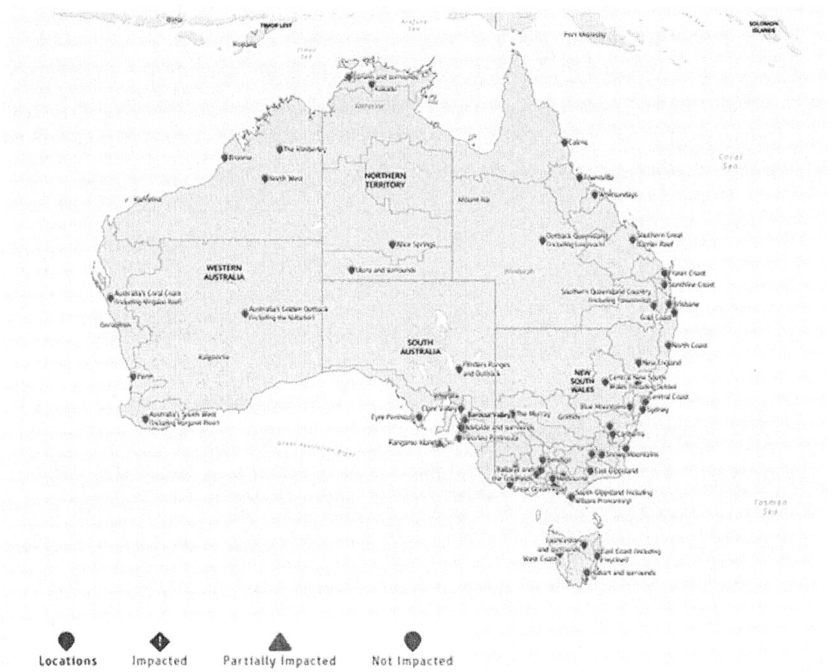

Figure 6.4 The extent of major bushfires in Australia during September 2019–January 2020, the largest bushfire outbreak in Australia since European settlement in 1788

Source: Tourism Australia, 2020. Used with permission

(Continued)

All emergency management agencies worked closely to aid in rescue and evacuation operations, including the military. Australian naval vessels evacuated hundreds of people from coastal communities in southern and North-East Victoria in early January 2020. Many hotels and caravan parks served as accommodation for bushfire fighters. In many respects, the concepts involved in the integration of emergency management and tourism were implemented during the Australian bushfires (Beirman, 2011).

As discussed in Chapter 3, Tourism Australia and the state tourism boards of New South Wales, the Australian Capital Territory, Victoria and South Australia and local tourism boards in those states actively promoted solidarity tourism to those communities, which were prepared and willing to accept visitation after the fires.

The Australian government approved a royal commission to examine risk, crisis and the management problems associated with the 2019–2020 bushfires. The full report was released in October 2020 (Australian Government, 2020).

Bush or forest fires are not unique to Australia. In recent years, major forest fires have caused extensive environmental and property damage in areas frequented by tourists in the USA, France, Spain, Italy, Israel and Greece. As human populations expand into forested areas, which in recent years have experienced all-time record temperatures and droughts, bush and forest fires are likely to become a growing challenge. As this chapter was being written (late 2020), massive and destructive forest fire destroyed vast areas of forest and property throughout the western states of the USA in California, Oregon and Washington state.

SUMMARY AND CONCLUSION

This chapter has sought to make the point that natural disasters are ubiquitous and can strike any location and any enterprise. For tourism destinations, risk management preparation for those natural disasters most likely to occur is vital. In some parts of the world the threat of climate-related natural disasters is a high and frequently recurring risk. The temptation to bow to market demands and build over-sea resort accommodation at a coastal resort may indeed attract many guests when the weather is kind. However, in the rare instance of a sea surge, hurricane or tsunami, this luxurious accommodation could be a death trap.

The most effective recovery programmes for destinations and tourism enterprises following a natural disaster focus on applying the lessons and noting the vulnerabilities experienced during past disasters, anywhere in the world, and applying them to measures that minimise exposure to future natural disasters; in other words, building back better.

The random nature of natural disasters elicits an unusually high degree of sympathy and empathy towards destination communities affected by them. Several of the case studies and examples referred to in this chapter illustrated marketing recovery strategies, which leverage off this genuine human response to natural disasters. Many tourists genuinely want to help a destination for which they have an affinity by supporting its recovery after a natural disaster. Provided the destinations affected by a natural disaster are willing to welcome tourists and view them as making a contribution to recovery, then destination marketers and tourism enterprises should communicate the message to the appropriate markets.

Discussion Questions and Issues

1. In reference to a natural disaster of your choice which has impacted on tourism, how would you apply the principle of building back better to the recovery of tourist visitation and tourism infrastructure?
2. Solidarity tourism applies to the recovery of many destinations affected by natural disasters. In reference to a natural disaster of your choice, how is solidarity tourism incorporated in the recovery of a destination or tourism enterprise and what sort of people would be targeted for such recovery marketing campaigns?
3. Develop a risk management matrix for a destination or a tourism business of your choice with a focus on the likelihood and consequence of natural disaster risk, and discuss how you would choose to treat those.
4. In reference to a natural disaster of your choice, discuss how and if volunteer tourism is appropriately incorporated in a post-natural disaster recovery campaign.

REFERENCES

Australian Government (2020) *The Royal Commission into Natural Disaster with Emphasis on the Australian Bushfire Crisis 2019–2020*. https://naturaldisaster.royalcommission.gov.au. Accessed 8 December 2020.

BBC News (2018) Lombok Earthquake and Tsunami. https://bbc.com/news/world-asia-45238018. Accessed 7 December 2020.

BBC News (2020) *Australian Fires. A Visual Guide to the Bushfire Crisis*. https://bbc.com/news/world-australia-50951043. Accessed 8 December 2020.

Beirman, D. (2003) *Restoring Tourism Destinations in Crisis. A Strategic Marketing Approach*. Allen & Unwin: Crows Nest (Sydney), Australia.

Beirman, D. (2011) 'The integration of emergency management and tourism', *The Australian Journal of Emergency Management*, 26(3): 30–34.

Beirman, D. (2018) 'Thailand's approach to destination resilience: An historical perspective of tourism resilience from 2022 to 2018', *Tourism Review International*, 22(3–4): 277–92.

Beirman, D., Upadhayaya, P., Pradhananga, P. and Darcy, S. (2018) 'Nepal earthquake in the aftermath of the April/May 2015 earthquake and aftershocks. Repercussion, recovery and the rise of new tourism sectors', *Tourism Recreation Research*, 43(4): 1–11.

Beirman, D., Wearing, S., Fullerton, S. and Darby, M. (2019) *The Development of a Destination Resilience and Tourism Recovery Strategy: Tathra's Response to Bushfires of March 2018*. Tathra Chamber of Commerce: Australia.

Benton, N. (2019) 'Island Tragedy', *Australian Leisure Management*, 137: 56–7.

Blakely, E. (2011) *My Storm. Managing the Recovery of New Orleans in the Wake of Katrina*. University of Pennsylvania Press: USA.

Flynn, C. (2017) Personal communication between the author and Chris Flynn, who related the post tsunami conference in Samoa when we were both speakers at the Pacific Insights Tourism Conference. Port Vila, Vanuatu 25 October, 2017.

Fullerton, S. and Darby, M. (2019) *Tathra, It's in our Nature. A Community's Experience of a Bushfire: Impact and Recovery*. Tathra & District Business Chamber: Tathra, Australia.

Global Disaster Alert and Coordination System (2020) www.gdacs.org. Accessed 7 December 2020.

Government of Nepal (2019) *Nepal Tourism Statistics 2019*. Singha Durba, Kathmandu, Ministry of Culture, Tourism and Civil Aviation, Planning and Evaluation Division, Research and Statistical Section: Nepal.

Gurtner, Y. (2007) 'Phuket: tsunami and tourism – a preliminary investigation', in Laws, E., Prideaux, B. and Chon, K. (eds), *Crisis Management in Tourism*. Cabi Publishing: Wallingford, Oxon, UK. pp. 234–51.

Kharzai, B., Mahdavian, F. and Platt, S. (2018) 'Tourism recovery scorecard. Benchmarking and monitoring progress on disaster recovery in tourism destinations', *International Journal of Disaster Risk Reduction*, 27: 75–84.

McDonald, F. (2005) *Presentation on the Meteorological Changes and Altered Hurricane Paths*. Best Education Network, June 2005, Think Tank 3: Kingston, Jamaica.

Pacific Asia Travel Association (PATA) (2015) *The PATA Nepal Tourism Rapid Recovery Taskforce Report and Recommendations*. PATA: Bangkok. This can be viewed on the PATA website, www.pata.org.

Pararas-Carayannis, G. (2009) *Earthquake and Tsunami of 29 September 2009 in the Samoan Islands*. https://www.drgeorgepc.com/Tsunami2009Samoa.html. Accessed 10 May 2021.

Prideaux, B., Coglan, A. and Falco-Mammon, F. (2008) 'Post crisis recovery. The case after cyclone Larry', *Journal of Travel and Tourism Marketing*, 23(2–4): 163–74.

Takamatsu, M. (2014) 'The Okinawa tourism crisis management initiative', *International Journal of Event Management Research*, 8(1): 19–34.

Turner, L. and Ash, J. (1975) *The Golden Hordes and the Pleasure Periphery*. Constable and Co.: London, UK.

UN Dispatch (2015) *Nepal Earthquake Facts and Figures*. https://www.undispatch.com/nepal-earthquake-facts-and-figures/. Accessed 7 December 2020.

UN World Tourism Organisation (UNWTO) (2020) *Tourism Emergency Response Network*. https://unwto.org/esilience-tourism-development. Accessed 7 December 2020.

Upadhayaya, P. (2018) 'Sustainable management of trekking trails for adventure tourism in mountains. A study of Nepal's Great Himalaya Trail', *Journal of Tourism and Adventure*, 1(1): 1–31.

Van Walbeek, B. and The Winning Edge (2003) *Crisis: It Won't Happen to Us!* PATA: Bangkok.

Victorian Government Tourism Resources (2020) https://business.vic.gov.au/tourism-industry-resources. Accessed 8 December 2020.

Wearing, S., Beirman, D. and Grabowski, S. (2020) 'Engaging volunteer tourism in post-disaster recovery in Nepal', *Annals of Tourism Research*, 80: 102802.

Weaver, D. and Lawton, L. (2014) *Tourism Management* (fifth edition), John Wiley and Sons: Milton, Queensland, Australia. pp. 83–9.

7

THEME 4: HEALTH-RELATED TOURISM CRISES WITH EMPHASIS ON PANDEMICS

INTRODUCTION

Concerns related to health issues and diseases have long been a factor in destination decision making for travellers. They have also been, and remain, a consideration for travelling in the present day. During 2020, Covid-19 dominated the global tourism to a greater extent than any pandemic in recent history. As Chapter 3 of this book is devoted to discussing Covid-19, this chapter examines the broader issue of health-related risks and crises, which arose prior to 2020, and focuses on three pandemic outbreaks prior to 2020: severe acute respiratory syndrome (SARS) in 2003, H1N1 (swine flu) in 2009–2010 and Zika during 2005–2017. These pandemic outbreaks had a significant impact on global tourism and spurred a global tourism industry response – supported by individual destinations and tourism businesses.

Many older travellers bear the scars of a smallpox vaccination, a longstanding requirement for international travellers until the World Health Organisation (WHO) declared the eradication of the disease in 1980. The WHO maintains a list of required vaccinations for specific destinations to protect travellers from diseases endemic to these countries, including cholera, typhoid and yellow fever. Travellers visiting affected destinations are required to produce proof of vaccinations when entering these countries and upon return to their home country. The quality, cost and extent of healthcare in destinations is frequently a factor in destination choice for international travellers, especially those with chronic or pre-existing health problems. Information on health risks is usually included in government travel advisories relating to intended travel destinations (WHO, 2020).

Quite a number of medical practitioners worldwide specialise in providing advice and preventative vaccinations and treatments for travellers who visit destinations with identified health risks. Some medical practitioners have formed associations such as the Travel Doctors and Traveller Medical Service to assist travellers in locating medical practitioners with expertise in travel medicine. Major travel insurance companies recommend medical practitioners to treat travellers in all countries covered on their policies. International SOS, one of the world's largest travel insurers for corporate clients, contracts medical and rescue teams for rescue and evacuation in remote regions or mountainous terrain. Travellers are not officially required to undertake vaccinations and drug treatments for the various strains of hepatitis, tetanus, malaria and other infectious diseases. However, preventative treatment of these diseases can be invaluable risk management practice in destinations where the risk of contracting these conditions is high (International SOS, 2020).

ACCESSIBLE TOURISM

Closely related to health risks is the issue of accessible tourism. A surprisingly large percentage of almost ten per cent of international travellers are either permanently or temporarily disabled (Smith, 2013). These disabilities range from blindness, deafness, temporary mobility restrictions due to injury, sudden illness or permanent mobility restrictions due to paraplegia, quadriplegia, amputations or degenerative diseases. Destinations with the tourism infrastructure to cater effectively to accessible tourism have access to a very significant market sector.

The United Nations (UN) declared 1981 as the Year of Disabled Persons. Consequently, many sectors of the tourism industry led by the accommodation sector, airlines, rail and coach operators, taxis and hire cars upgraded facilities in many countries to enhance access and support for temporarily and permanently disabled travellers. In 2016, the UN World Tourism Organisation (UNWTO) released accessible tourism guidelines for the global tourism industry, which were regularly updated (UNWTO Accessibility, 2020). Academic research in the field of accessible tourism has been led by Professor Simon Darcy from the University of Technology Sydney (UTS) and Professor Dimitri Buhalis from Bournemouth University (Buhalis and Darcy, 2010) both of whom contributed their expertise to the UNWTO's accessible tourism guidelines.

A growing number of tour operators worldwide specialise in operating accessible tour programmes. In Nepal, Mr Pankaj Pradhananga is CEO of Four Seasons Travel, which has pioneered and advocated accessible tourism in Nepal. In common with many developing countries, accessible tourism infrastructure in Nepal is limited but Four Seasons Travel has succeeded in gaining the cooperation of many tourism companies and associations to ensure that groups of disabled visitors can enjoy a full tourism experience in Nepal. After the 2015 Nepal earthquake, Mr Pradhananga and some of his Nepalese tourism colleagues developed a wheelchair accessible trail for people to enjoy the vistas of the Annapurna Range, one of Nepal's most spectacular sites. The development of accessible infrastructure has contributed to Nepal's post-earthquake recovery (Beirman et al., 2018).

Israel is a world leader in catering to the needs of disabled travellers or those with a need for health treatment of chronic illnesses while visiting the country. Given Israel's history of war and conflict and since its establishment in 1948, and the large number of Israelis who have sustained disabling injuries, Israelis are very conscious of treating and catering to the injured and disabled. Yad Sarah (2020) was the world's first volunteer-based organisation providing special health treatment for locals and tourists in Israel who require treatment and equipment for a wide range of both chronic conditions or treatment in response to an injury incurred while in Israel. They also loan, free of charge, wheelchairs and other mobility aides and prosthetic devices to travellers and locals who may need them in Israel. Yad Sarah was established in 1976 by Rabbi Dr Uri Lupolianski (Mayor of Jerusalem 2003–2008) and currently has 100 centres all over Israel. It remains one of the few institutions of its type to offer free medical support for tourists with ongoing disabilities or who require support and medical equipment following injury. As Israel attracts a large number of religious pilgrims of all faiths, many of whom are elderly or infirm, the services provided by Yad Sarah are a best practice model for many countries who seek to provide destination-based accessible travel services.

TOURISM INDUSTRY SECTORS AND MEDICAL RISK MANAGEMENT

Most tourism industry sectors are highly aware of medical risks. From a risk management perspective, healthcare preparedness is an integral part of their duty of care for their employees, guests, participants or passengers. Under occupational health and safety laws, the commitment to health and safety applies equally to staff and customers. Cruise ships in effect are communities at sea and routinely have fully-staffed medical operations, including qualified medical practitioners, nurses and on larger cruises, full hospital facilities. Cruise ship and airline crew are required to undertake first aid training and achieve proficiency. As a significant cohort of cruise passengers are elderly and frail, attention to medical care is taken very seriously by cruise ship operators and is an important part of the Saving of Life at Sea (SOLAS) protocols of the International Maritime Organization (IMO) (SOLAS, 2020). Although risk management in the cruise sector is highly regulated globally through the IMO and the cruising sector's global association, Cruise Lines International Association (CLIA), cruising does have some obvious vulnerabilities from a health perspective. Cruise ship crews take special care with storage, preparation and serving of food and beverage but occasionally episodes of food poisoning do occur. While cruise ships are required to have full-time doctors, nursing and medical staff and facilities (including isolation cabins), this does not protect cruise passengers from contracting infectious diseases. Cruising's great attraction as a highly sociable form of travel has also exposed its vulnerability to epidemics and pandemics.

Although airlines are not required to carry a qualified medical practitioner on flights, all cabin crew are trained in emergency and first-aid medical treatment. Qualified medical practitioners who fly as passengers are requested to make themselves known to aircraft crew. Most major airports include a full range of medical services and frequently engage in health screening of passengers who may be arriving from an at-risk country or during pandemics.

Many tour operators, especially those tours emphasising active or adventure travel components, require tour participants to provide evidence that they are medically fit to participate in tours. This is especially important for those adventure tours requiring a high level of physical fitness. Reputable adventure tour operators ensure that either a guide or a tour leader is qualified to provide first-aid treatment for participants who suffer injury or illness during a tour. Event organisers, especially those organising large-scale events, are legally required in some jurisdictions to have on-site medical services. Organisers of events such as 'fun runs' will have medical clinics at strategic points of the 'fun run' with a concentration just past the finish line.

The accommodation sector, especially large hotels, generally treat health threats seriously. Many larger hotels employ an in-house doctor or nurse and have ready access to a number of medical practitioners in the vicinity. One of the many roles of a hotel concierge is the ability to source medical treatment for guests who may require specialist or emergency medical attention. Many hotels set aside rooms with enhanced access and equipment, catering to guests with disabilities in addition to ramp access within the hotel and from the street to the hotel entrance.

A major element of risk management for international travellers is the purchase of travel insurance. Travel insurance is strongly recommended in most government travel advisories. The travel industry is encouraged to sell travel insurance and many travel insurance companies pay travel agents attractive commissions to do so. While the national medical insurance programmes of some countries include reciprocal coverage in selected foreign countries, this is the exception rather than the rule. One of the many uncertainties surrounding Brexit is whether British citizens covered by the UK's Global Health Insurance Card (GHIC) will have reciprocal coverage in European Union (EU) countries, as was the case when the UK was a member of the EU.

There are limited reciprocal medical arrangements between Australian and New Zealand citizens in their respective countries and with the UK's National Health Service (NHS). Most travel insurance companies provide unlimited medical coverage provided it's not associated with pre-existing medical conditions or in circumstances where the traveller's irresponsible or illegal behaviour or inebriation was the cause of the injury or medical condition.

A small minority of international travellers erroneously believe that if they encounter medical difficulties in a foreign country and are uninsured, their government will come to their aid, pay their medical expenses and repatriate them. On very rare and exceptional occasions, governments may do this for their citizens but the financial cost will involve a loan repaid with interest. As many officials in Australia's Department of Foreign Affairs and Trade have repeated in over 35 Consular Consulting Group meetings since 2003 (Australian DFAT, 2020), it is not the government's role to subsidise the ignorance and stupidity of irresponsible citizens travelling abroad. For holiday travellers especially, travel is a discretionary activity and the traveller is ultimately responsible for insurance coverage.

THE CHAPTER'S FOCUS

This chapter examines the issue of health-related risk, crisis and recovery largely from the destination perspective. The perception or reality that a destination region contains high risks for visitors contracting certain diseases is an influential factor for destination choice. The vulnerability of and risks to tourists of disease is also a factor in choosing the mode of travel. As discussed in Chapter 3, which focused on Covid-19, risks associated with cruising, events, conventions, large group tours and other sociable forms of tourism aroused a significant fear factor.

This chapter also discusses the broader issue of health-related tourism concerns and pandemics – apart from Covid-19. Pandemics have figured prominently in tourism crises during the first 20 years of the 21st century. SARS, H1N1, HIV AIDS, avian flu, MERS, Zika, measles and Ebola are just some of the more headline-grabbing pandemics impacting on tourism since 2000. There are many more diseases, which have affected tourists less heavily featured by the global media.

As discussed in Chapter 3, Covid-19 posed a range of new challenges to the viability of tourism destinations and enterprises. However, some of these new responses build on the strategies that worked to recover from earlier health risks and pandemics. The case studies focus on three significant pandemic outbreaks in recent years. All three pandemics involved very different approaches to managing the tourism impacts and in restoring tourism confidence to affected destinations.

DEFINING KEY HEALTH AND MEDICAL TREATMENT RISKS FOR TOURISM AND PANDEMICS

When travellers choose a destination, safety in all its forms is a key concern. Integral to safety is the issue of health risks. Many destinations in Africa and Latin America require travellers to show evidence that they have had recent vaccinations against yellow fever, cholera and typhoid stamped in their yellow WHO vaccination document (see Figure 7.1). This evidence is frequently a re-entry requirement for travellers who have visited countries affected by these diseases on return to their home country. International travellers are not legally required to undergo vaccinations against a wide range of diseases. However, medical practitioners, especially those who specialise in travel medicine, often advise a course of preventative vaccinations for travellers visiting 'at risk' destinations. These treatments may include preventative measures against hepatitis, malaria and tetanus – among many others. Increasingly, medical practitioners provide advice on dietary practices and protective clothing that minimise the risk of exposure to a range of illnesses.

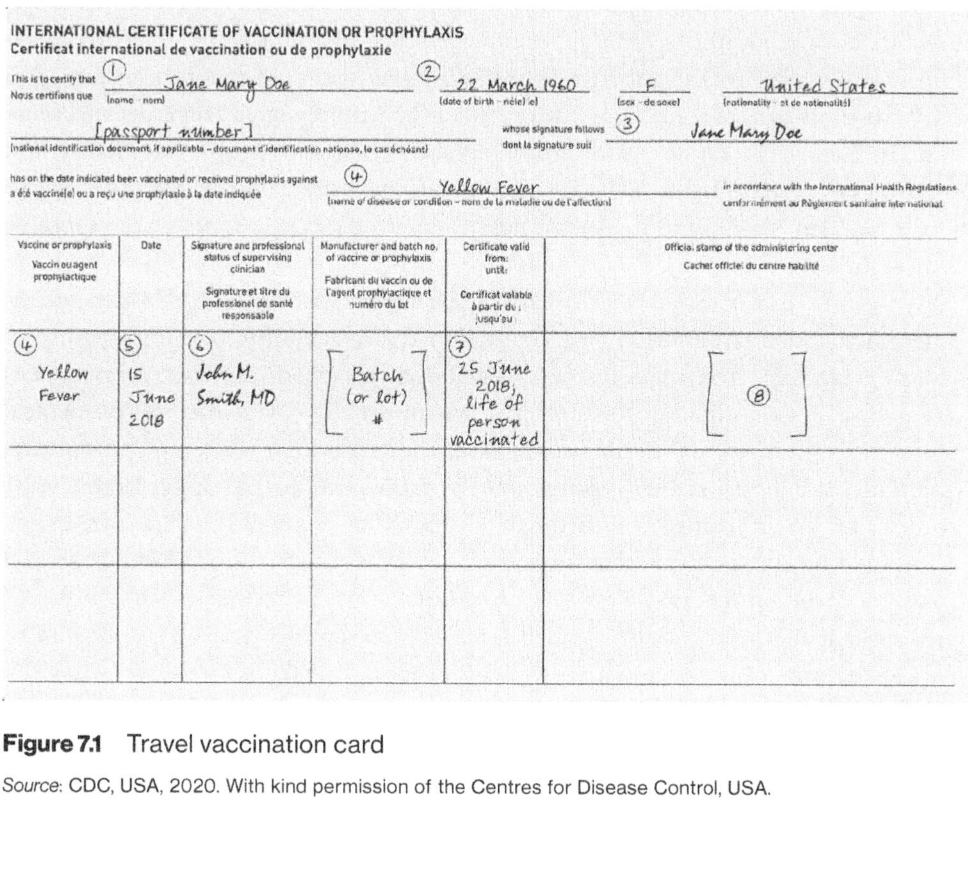

INTERNATIONAL CERTIFICATE OF VACCINATION OR PROPHYLAXIS
Certificat international de vaccination ou de prophylaxie

This is to certify that / Nous certifions que ① Jane Mary Doe (name – nom) ② 22 March 1960 (date of birth – né(e) le) F (sex – de sexe) United States (nationality – et de nationalité)

[passport number] (national identification document, if applicable – document d'identification national, le cas échéant) whose signature follows / dont la signature suit ③ Jane Mary Doe

has on the date indicated been vaccinated or received prophylaxis against / a été vacciné(e) ou a reçu une prophylaxie à la date indiquée ④ Yellow Fever (name of disease or condition – nom de la maladie ou de l'affection) in accordance with the International Health Regulations / conformément au Règlement sanitaire international

Vaccine or prophylaxis Vaccin ou agent prophylactique	Date	Signature and professional status of supervising clinician Signature et titre du professionel de santé responsable	Manufacturer and batch no. of vaccine or prophylaxis Fabricant du vaccin ou de l'agent prophylactique et numéro du lot	Certificate valid from: until: Certificat valable à partir du : jusqu'au :	Official stamp of the administering center Cachet officiel du centre habilité
④ Yellow Fever	⑤ 15 June 2018	⑥ John M. Smith, MD	[Batch (or lot) #]	⑦ 25 June 2018; life of person vaccinated	[⑧]

Figure 7.1 Travel vaccination card

Source: CDC, USA, 2020. With kind permission of the Centres for Disease Control, USA.

In addition to recommending that travellers purchase travel insurance, government travel advisories generally include a section for prospective travellers, outlining the health risks that apply to specific destinations. In recent years, as travellers became increasingly attracted to more remote destinations, beyond the tourist trail, travellers have paid increasing attention to medical risks. Some diseases, result from high-risk behaviour, which increases the likelihood of transmission. Sharing of needles for intravenous drug use and engaging in unprotected sexual activity increase the likelihood of contracting HIV AIDS and a range of sexually transmitted diseases.

Other health risks for travellers may include illness resulting from bites and stings from venomous insects, snakes, spiders, land animals and sea creatures. Poor sanitation, unclean food storage, preparation and contaminated water pose health risks for unwary travellers in many parts of the world. While many local people may have developed immunity from some of these items, visitors may be vulnerable and suffer allergic reactions and food poisoning in reaction to unfamiliar foods and beverages.

Apart from diseases, risks to health and physical wellbeing can result from accidents and mishaps sustained during the course of travel. Travellers who engage in active and adventure travel activities, including hiking, skiing, cliff and mountain climbing, abseiling, bungee jumping, surfing, swimming, scuba diving, kayaking, rafting, parachuting and sky diving can experience severe injuries in the course of those activities. Many travellers sustain injuries in motor vehicle and motor cycle accidents and some countries have very lax rules about hiring out such vehicles to people unqualified and unlicensed to operate them. Even for non-adventurous travellers, accidents incurred in falls, sudden illness due to stroke, heart attack or food poisoning are pervasive medical threats.

Beyond the risks of acquiring diseases or sustaining serious injuries, a key health-related factor in destination choice centres on the availability, cost and quality of medical treatment. The majority of international travellers purchase travel insurance policies for the duration of their trip, which usually covers the cost of medical treatment in international destinations (except for treatment related to pre-existing conditions) and in most cases, insurers will cover the cost of repatriation to the home country and related treatment if required.

Travellers need to be aware that there are exceptions and exclusions related to travel insurance cover, which vary between providers. Many travel insurance companies deny cover if the injury results from the traveller being inebriated from alcohol or drugs at the time of the incident. Travellers engaging in risky activities or behaviour (beyond those specified in their insurance cover) are denied coverage. Most insurers exclude coverage for illness arising from pre-existing medical conditions or an epidemic/pandemic, known before the date of departure. Very few people would dispute that the fine print of a travel insurance policy is dull reading. However, for travellers taking a major trip, it should be required reading. Alternatively, travellers should seek professional advice about the medical coverage inclusions and exclusions applying to their travel insurance policy.

Health-related issues that attract heightened media attention are those related to epidemics and pandemics, which present a risk to all visitors to a specific destination or to multiple destinations. For clarification, an epidemic involves a significant outbreak of a communicable disease largely confined to a single country. There is always a possibility that an epidemic can spill across national borders. By contrast, pandemics are communicable diseases with large numbers of cases in many countries. International travel frequently, if unintentionally, facilitated the spreading of these diseases.

The three case studies in this chapter are:

- SARS in 2003
- H1N1 (swine flu) in 2009–2019
- Zika in Brazil 2015–2016.

As mentioned earlier, there have been many more pandemic outbreaks since 2000 than the three featured in this chapter's case studies. The response to pandemics including SARS, H1N1 and Zika involved a number of key stages:

1. Identification of the condition and its key symptoms.
2. Screening and isolation of affected patients at points of entry to transport providers and border controls.
3. Containment of the spread of the disease through quarantine, social distancing regulations and isolation of identified infected patients.
4. Treatment of the disease when identified. As we have already experienced with Covid-19, the development of a vaccine or medical treatment can involve a lengthy process of research and testing prior to manufacture and distribution of a treatment. Even when a vaccine or treatment regime has been developed, there is no guarantee it can treat the disease for all patients.
5. Ensure all relevant stakeholders are informed of measures taken in response to the outbreak and release relevant data on case numbers, recommended preventative measures and hotspots in the destination region. Effective crisis communications are vital to clearly distinguish perception and distortion from verifiable facts. Honest and trustworthy engagement with the media is critical in the management of all tourism-related crises but as pandemics often involve invisible elements (viruses) that arouse panic, factual engagement with the media on all its platforms, backed by expert medical authorities, is required.
6. Restoring consumer, stakeholder and media confidence that the disease is sufficiently controlled in a given destination region to re-market the destination.

Case Study 7.1

SARS, 2003

SARS, in common with Covid-19, was a virulent variant of coronavirus. This condition, known as severe acute respiratory syndrome emerged in southern China in early 2003 and spread into Chinese communities in Hong Kong, Singapore and many countries in South-East Asia. A cluster, which also originated in China, occurred in Toronto, Canada. The mortality rate of SARS at ten per cent was much higher than Covid-19 (2.5 per cent).

In the scheme of pandemics, SARS was a relatively small and effectively contained outbreak. According to the WHO, during 2003 just over 8,000 people were affected by SARS of whom 800 died (WHO, 2004). Even by 2020 there has never been a cure or a vaccine developed to treat SARS. The first SARS cases were recorded in November 2002. By July 2003, it had virtually disappeared. The only treatment that worked was the isolation of people who had contracted the disease. Yet during that short period, SARS and some of the lurid media coverage of it created fear and panic among travellers, especially in the Asia Pacific region and in Canada. SARS was a classic example of a perceptual health crisis and the perception of its threat to health led to sudden and significant declines in tourism in countries where there were a substantial number of cases, notably in China, Hong Kong and Singapore. Other South-East Asian countries registered a small number of cases, notably Thailand and the Philippines (Wilder-Smith, 2006; Beirman, 2009), but in these countries the perception of a health threat led to steep short-term declines in international tourism arrivals even in countries with low case levels. Figure 7.2 illustrates the main countries in which cases occurred.

Country	Number of cases	Mortality rate number and percentage	First case date	Last case date
Canada	251	41, 17%	3 April	12 June
China PRC	5,327	349, 7%	16 Nov 02	25 June 03
China Hong Kong SAR	1,755	300, 17%	15 Feb	31 May
China ROC/Taiwan	665	180, 27%	25 Feb	15 June
Philippines	14	2, 14%	25 Feb	5 May
ROK/South Korea	3	0, 0%	25 April	12 May
Singapore	238	33, 14%	25 Feb	5 May
Thailand	9	2, 22%	11 March	27 May
UK	4	0, 0%	1 March	1 April
USA	33	0, 0%	9 Jan	13 July
Vietnam	63	5, 8%	23 Feb	14 April
WORLD	8,422	916, 11%	16 Nov 02	13 July 03

Figure 7.2 Cases of SARS in 2003, selected countries

Source: WHO, 2004. Figure by the author

Although SARS was a communicable disease, it was largely transmitted through contact between bodily fluids. SARS was not an airborne spread disease and this enabled the isolation of patients to be an effective treatment (WHO, 2004). During the 2003 SARS outbreak IATA insisted on enhanced airline air conditioning filtration units installed in all member airlines. These were designed to kill both bacteria and viruses.

By comparison with other pandemics, SARS was a relatively small outbreak. The vast majority of cases (7,985 out of 8,422 cases globally) were in five jurisdictions: China, Hong Kong, Taiwan, Canada and Singapore. In all cases, those affected were overwhelmingly people of Chinese origin. Canada was the only country outside of Asia with a significant number of cases, most of them transmitted within Toronto's Chinese community.

Despite the above facts, SARS aroused a level of fear far in excess of its reach and severity. Tourists were literally terrified to travel anywhere within South-East Asia. Media images depicted commuters in Hong Kong and Singapore going to work and shopping while wearing surgical masks. Even the normally sober media outlets depicted SARS as a rampant plague endangering travellers. The global media depicted SARS as a major pandemic threat to Asian tourism. International tourism visitation to Hong Kong and Singapore plummeted between April and July 2003 compared to the previous year. In the case of Singapore, visitation plummeted by 70 per cent between March and June 2003 compared to the 2002 levels and there was a similar decline in Hong Kong (Henderson, 2003). Significant declines occurred in China and Vietnam. This may have been partially justified as there were a number of cases in both Hong Kong and Singapore. However, during the second quarter of 2003 (April–June) tourists virtually abandoned countries such as Thailand, Malaysia and the Philippines, which between them had 28 cases during the entire SARS outbreak.

Wilder-Smith (2006) claimed that in April 2003 tourism to South-East Asia declined by 41 per cent compared to the corresponding month in 2002. Collectively, tourism to the Asia Pacific region between April and June 2003 declined by 12 million arrivals compared to the corresponding period in 2002.

Even countries outside the South-East Asian region such as Australia and the South-West Pacific, India and Europe were affected by SARS as many travellers avoided even transiting via South-East Asia to reach Europe or North America. The international media, and to a lesser extent the WHO, aroused an almost irrational sense of fear and foreboding about the likelihood of traveller exposure to SARS.

The Pacific Asia Travel Association (PATA) represented all countries affected by SARS, including Canada. In cooperation with Association of Southeast Asian Nations (ASEAN) tourism and with the support of Tourism Industry Association of Canada, PATA coordinated an Asia Pacific tourism recovery campaign. The campaign, referred to as Project Phoenix, emerged as a template for many subsequent tourism recovery campaigns. PATA's CEO at that time was Peter de Jong who proved to be a highly effective spokesman for PATA's campaign.

(Continued)

In a TV interview with CNBC in April 2003, de Jong (Hospitality Net, 2003) summarised the intent of Project Phoenix. He described it as a reputation management campaign designed to restore consumer confidence in the Asia Pacific as a desirable destination region. The driving force of Project Phoenix was the need to counter negative and exaggerated perceptions about SARS with factual evidence to prove that tourists were both welcome and safe in the Asia Pacific region. PATA would act as the single voice for the region but relied on all each Asia Pacific countries to promote and market their own destination area independently. The strategy's core premise focused on motivating tourists to visit the region. De Jong pointed out that the campaign would 'plough the field' to restore confidence and that each participating destination and business would 'sow the seeds' with incentives, coupled with destination and tour product promotions offered by each destination and tourism business, to encourage tourists to return to the region (Flynn, 2020).

In essence, this was a collaborative destination marketing campaign. The core collaborators were media outlets, including CNN and BBC, tourism Ministries from all affected countries, airlines, tour operators, cruises companies, accommodation providers and travel agents. PATA, as a transnational tourism association whose membership included both government tourism ministries and private sector firms, was in a unique position to direct this programme. Advertising and promotion employed all available media platforms and celebrities for the various countries to contribute their own positive messages. As the literature has indicated (Beirman, 2009; Henderson, 2003; Yates, 2006), Project Phoenix was an outstanding success, aided largely by the fact that SARS cases dwindled after June 2003. By the end of 2003, tourism to all affected destinations had recovered but the cost of a perceptual health crisis was steep for the Asia Pacific travel region. The campaign became a template for multiple tourism recoveries for years after 2003, until Covid-19 required a re-writing of tourism recovery campaigns.

THE MEDIA AND PANDEMICS

A unique aspect of PATA's Project Phoenix was the extensive involvement of major global media outlets in PATA's recovery campaign. CNN and BBC were two of the most prominent media organisations affiliated with PATA in 2003. Lest anyone think that the BBC or CNN provided significant free air time to spread recovery messages about the tourism to the Asia Pacific region because they felt any charitable impulses, this was not the case. BBC and CNN's support for Project Phoenix occurred as matter of primal self-interest. They and many other media outlets contributed to spreading the fear and the panic about SARS. CNN and BBC delivered a quiet but cogent message by their many tourism industry stakeholders that their heavy reliance on tourism advertising and in-kind support involved a quid pro quo. In other words, trashing the tourism industry is literally biting the hand that feeds many media outlets. The global tourism industry, especially in relation to SARS, had every right to play hardball with the media.

Former US President Donald Trump was often criticised for exaggerating his contempt towards the media. However, his term 'fake news' contained a ring of truth in relation to the media's coverage of SARS and many other pandemics discussed in this chapter.

Media outlets elicit a great deal of assistance from tourism organisations, airlines, tour operators and hotels to aid them, news, documentaries, entertainment and travel shows. CNN, BBC and media outlets communicating on traditional and online platforms also receive vast amounts of advertising revenue from tourism destination marketing authorities and private sector tourism and hospitality business. SARS, in common with the early phase of H1N1 and to a lesser extent Zika, contain an element of media-driven hysteria elevating negative perceptions about destinations far in excess of the objective level of threat.

Eli Avraham and Eran Ketter are two of the world's leading researchers in the field of media strategies for marketing places in crisis (Avraham and Ketter, 2008). Pandemics and epidemics are one of the five key crisis types for which tourism destination and business marketers should have a media strategy. The other four are crime, terrorism, political unrest and natural disasters. In order to counter negative perceptions, the counter-argument needs to be factual, verifiable and compelling. Central to any effective crisis communication or post-crisis recovery media campaign is the importance of working with the media as a partner, not as an adversary (Nielsen, 2001).

Most national tourism organisations have a visiting journalist programme based on the golden rule of media marketing, 'don't blow your own trumpet, get a credible person to blow it for you'. There is considerable value in dealing with negative perceptions, particularly the impact of a pandemic to enable the media to see the facts for themselves and communicate them in their own words rather than being asked to read from a script. Destinations with a compelling case, truthfully contradicting negative perceptions, can usually rely on media professionals to report the story without direction. The rider to this is that some media outlets and programmes do twist a positive story into a negative one. Destination and tourism managers need to be judicious about journalists, media outlets and programmes they choose to work with.

Case Study 7.2

H1N1 (Swine Flu), 2009–2010

By 8 November 2020, the global number of Covid-19 cases exceeded the 50 million mark and deaths attributed to the pandemic exceeded 1.25 million, representing a 2.5 per cent mortality rate.

(Continued)

H1N1, better known as swine flu, represented a coronavirus threat to much of the world (213 countries) between April 2009 and April 2010 and beyond. The disease involved flu-like symptoms communicable by air droplets via sneezing. The number of people affected by swine flu varied greatly according to the sources consulted. The Centre for Disease Control (CDC) and the WHO claimed that 60.8 million people were officially diagnosed with swine flu (CDC, 2010). A University of Minnesota study, conducted under the auspices of the Center for Infectious Disease Policy and Research (Roos, 2011), claimed that based on the observation that many patients were asymptomatic, between 11 and 17 per cent of the world's population at the time (6.7 billion) were infected, representing a number between 700 million to 1.2 billion people.

The major distinction between H1N1 and Covid-19 was the mortality rate. The CDC estimated that between 151,700 and 575,000 died as a result of H1N1. Based on the CDC estimates, this represents a mortality rate of between 0.25 per cent and 0.9 per cent. In either event, the mortality rate of H1N1 was equivalent to standard influenza and certainly far lower than Covid-19. The main reason for the uncertainly over death rates was that most people who were reported as dying from H1N1 were people who were suffering from other medical conditions. Those most seriously affected by H1N1 were elderly, infirm or affected by other medical conditions, which made the diagnosis of cause of death problematic.

When the initial outbreak of H1N1 occurred in Mexico in March 2009, there was a media frenzy that erupted from the US on the dangers of this newly discovered disease (CNN, New York Times, CNBC, 2009). There were countless images of Mexicans wearing surgical masks and there was a fear generated that this could be the new SARS. This hysteria rapidly moderated when it was realised that H1N1 was a condition with a low mortality rate, equivalent to standard forms of influenza. In common with Covid-19, the WHO was slow to declare H1N1 as a pandemic. This occurred in June 2009 (WHO, 2009).

H1N1 ultimately appeared in 213 countries. While its low mortality rate reduced the fear associated with the virus, the development of a vaccine introduced in December 2009, which successfully treated 60 per cent of cases by aiding antibodies, further reduced H1N1's impact.

The global travel industry, using PATA's Project Phoenix (Beirman 2009) as a template, communicated the message that H1N1, while a hazard, was well short of a crisis for tourism and emphasised preventative measure that travellers could adopt to minimise their exposure to the virus. With the notable exception of Mexico, H1N1, despite the large number of people exposed to it, was not a global tourism crisis.

Under-publicised but severe health threats to tourists

There remains a wide range of ongoing health risks all over the world. Millions of people annually die from diseases, including malaria, HIV AIDS, typhoid, cholera, measles, sleeping sickness and a range of other maladies. These diseases rarely attract the prominent media coverage or widespread impact on tourism of Covid-19 discussed in Chapter 3, or other recent pandemic outbreaks discussed in this chapter. However, there are many diseases, some of which are highly contagious, which cause millions of deaths around the world annually and present a danger to tourists and tourism. The fact that the media largely ignores these diseases and their effects doesn't diminish the very real health threats they pose. However, from a perceptual perspective they have a low risk profile. It could be argued that the high mortality rates of the above diseases are concentrated in the developing world and consequently escape the attention of media outlets, which are predominantly based in the developed world.

Case Study 7.3

Zika virus and its Impact on the Rio 2016 Olympic Games, Brazil

The Zika virus is a mosquito borne disease, first observed in the Zika Forest, Uganda in 1947 (D' Olivera et al., 2017). Unlike SARS or H1N1, which arrived and disappeared within the space of eight months and two years respectively, Zika remains and continues to be active in some parts of the world. By July 2019, there was evidence of past and present transmissions of Zika in 87 countries and territories (WHO, 2019). Zika is prevalent in many tropical and sub-tropical regions of the world, especially forested and moist areas where mosquitoes thrive.

Although Zika is a disease that rarely kills people, it directly infects people with mild cold-like symptoms. Zika's primary danger is that its presence in an infected person, especially pregnant women, can give rise to birth defects. The birth defect causing the greatest medical concern is microcephaly, a condition in which a baby is born with a much smaller-than-normal-sized head, and this condition has a range of associated medical and mental development complications. In addition to microcephaly, Zika can cause adverse pregnancy outcomes, including stillbirth, limb contractures, eye abnormalities and brain calcifications (WHO, 2016).

The relevance of Zika to this chapter is that Brazil experienced the world's largest ever outbreak of Zika during 2015–2016, coinciding with the onset and running of the 2016 Rio Olympic Games (D'Olivera et al., 2017). In 2015 and 2016, there were 1,773,272 cases of Zika reported in Brazil of whom 2.5 per cent or 41,473 were pregnant women. Of these, there were 1,950 cases of babies born with microcephaly.

(Continued)

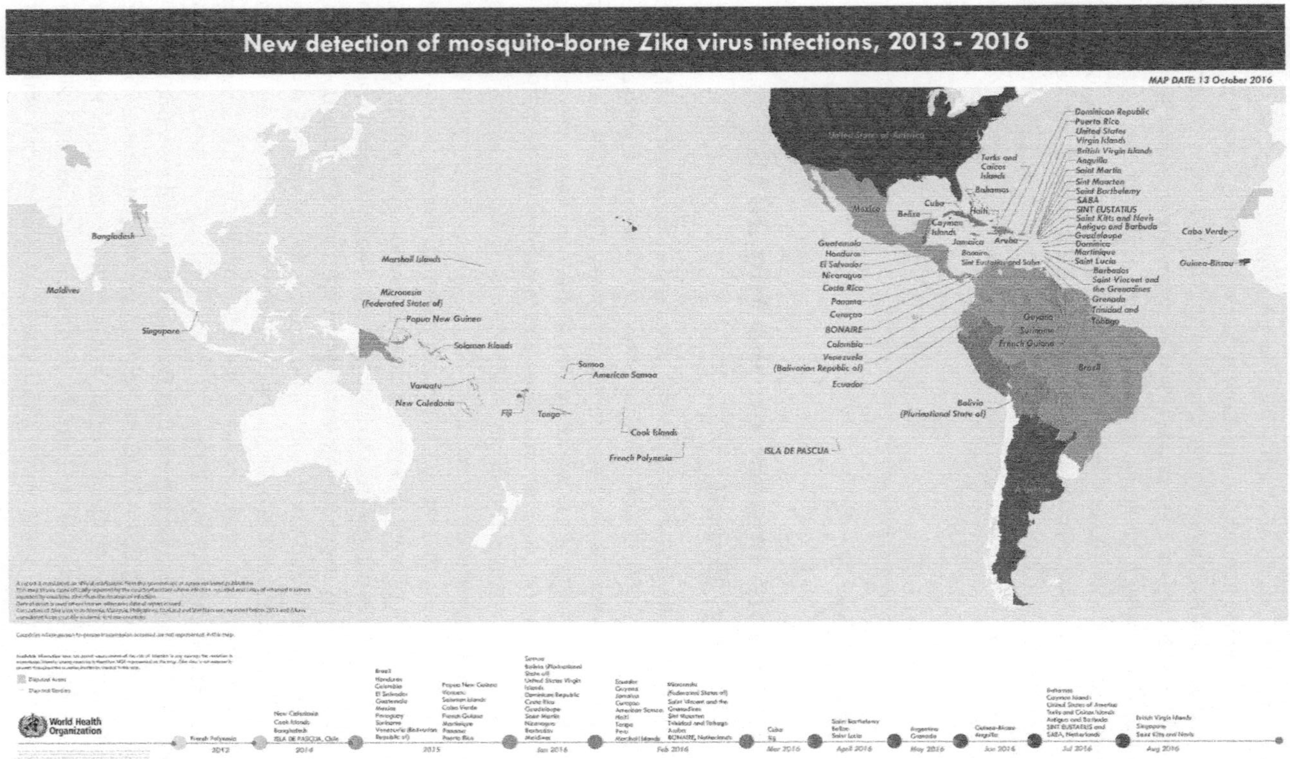

Figure 7.3 Zika's spread in 2013–2016

Source: WHO, 2016. With kind permission, WHO 2021. Reproduced from 'New detection of mosquito-borne Zika virus infections, 2013–2016 (2016).

During the lead-up to the Rio Olympics in August 2016, Brazil hosted literally thousands of media professionals encompassing all media platforms. Prior to the games themselves there were three major news items, which held the attention of the media: Brazil's high crime rate (discussed in detail in Chapter 8), social inequalities and associated anti-Olympic Games protests, and Brazil's Zika outbreak. As the birth defects associated with Zika are horrific, the more salacious elements on the media had a field day with horror images that frightened their audiences.

Although the objective risk of being infected with Zika was relatively small, the fear factor it aroused, especially among women of childbearing age, was high. There were a number of media reports focusing on female athletes from several countries who withdrew from competing at the Rio games, fearful of the risk of contracting Zika. Just before the Olympic Games, a letter signed by 200 academics from the USA and Canada called on the International Olympic Committee to cancel the Rio Olympics (Lewnard et al., 2016). The majority of Zika cases in Brazil were well over 1,000 kilometres north of Rio, concentrated along the Amazon basin in the north of the country.

A University of California (Berkeley) study indicated that not a single tourist or participant in the Rio Games contracted Zika (Israel, 2017) although some residents of slums in Rio did. In response to the concerns over Zika, there was considerable cooperation between Brazil's national tourism office (Visit Brazil) and a number of key stakeholders. These included the International Olympic Committee, WHO, UN World Tourism Organisation (UNWTO), World Travel and Tourism Council (WTTC) and PATA to address and counter many of the concerns surrounding the possible impact of Zika on visitation to Brazil during the Olympic Games (Lewnard et al., 2016).

All these organisations sought to downplay the extent of the threat of Zika to women and in fact issued advice on clothing and behaviour to minimise exposure to Zika-carrying mosquitos. The campaign focused on the low level of threat in Rio de Janeiro at that time of the year. The date of the Olympics, August, was still winter in the Southern Hemisphere and the likelihood of clouds of Zika bearing mosquitoes was minute. However, the campaign by Brazil's tourism industry and its allies to counter negative media coverage never dismissed or ridiculed the understandable concerns of women who were in at-risk groups.

It is difficult to measure the statistical impact that Zika had on tourism arrivals to Brazil before, during and after the 2016 Rio Olympics. The Zika factor clearly had its most powerful perceptual impact on women of child-bearing age and especially pregnant women. As discussed in Chapter 8, the parallel concern over Rio Olympic visitations centred on crime, which was a concern for all potential visitors. As Avraham and Ketter (2016) clearly explain, medical concerns form a significant part of a destination's image for better or worse. When the media is forensically covering a destination, which always occurs before, during and after a mega event, it is vital to have a media strategy to address a significant risk.

SUMMARY AND CONCLUSION

Medical issues represent a significant risk factor for tourism destination and tourism enterprises. Many sectors of the tourism industry have established protocols, which were generally successful in readiness for medical threats. As Chapter 3 has discussed, most of those protocols were found wanting with the advent of Covid-19 in 2020.

Health issues are integral to the safety of tourism. Destination management and marketing organisations do have a duty of care (whether they realise it not) to ensure that visitors are aware of health concerns in a country and have the means to treat them. Many countries have made significant progress in catering to the growing accessible tourism market. However, anyone who breaks a leg in a ski accident will find that getting around many destinations in the world when mobility is impaired can be exceeding difficult. Imagine what it is like for people who have permanent mobility constraints.

The case studies referred to different types of pandemics and their real and perceived impacts on tourism. SARS was the classic perceptual crisis but real enough to those unfortunate 8,400 people who actually contracted the virus, of whom 800 died. However, the fear and panic aroused by SARS and the scare-mongering reporting of WHO (which should have known better) and the global media were way out of proportion to the objective threat of the disease.

PATA's Project Phoenix, which was developed in 2003, proved to be a highly effective crisis communication campaign to counter the hysteria created by the misreporting of SARS and it has become a template for many crisis recovery campaigns for years following. H1N1 infected a vast number of people but despite initial panic and hysteria when the disease broke out in Mexico in March 2009, it shifted rapidly from a crisis to a hazard. The combination of a low mortality rate, a vaccine developed in late 2009 and a global travel industry, which used variations of the Project Phoenix template, saw off H1N1.

Zika was essentially one of three elements that made the promotion of the 2016 Rio Olympics and tourism growth to Brazil a challenge. The others were crime and glaring social inequalities. Zika's medical threat was to the babies of women infected by the mosquito borne disease, making this a threat to one important segment of the tourism market. As it happened, Brazil experienced the world's largest outbreak of Zika at the time of the 2016 Olympics in the full glare of a media hungry to report bad news. The challenge for Brazil's tourism industry and numerous stakeholders who sought the success of the Games was to collaborate in delivering the facts of the real threat of Zika to visitors to Rio and the measures that women especially could take to minimise risk. Facts, collaboration and context are three key bywords in responding to negative perceptions associated with health.

Discussion Questions and Issues

1. What would be the key sources of information and preparation strategies related to travel health that travellers could use in preparing an international trip to a destination of your choice?
2. Pick a country or a tourism industry sector of your choice and discuss the facilities or lack of them to assist wheelchair-bound travellers.
3. Project Phoenix is mentioned several times in the chapter as a crisis communication template. Assess its strengths and its weaknesses in relation to Covid-19.
4. Imagine that you are the Marketing Manager of Visit Brazil. Outline the strategy you would employ to respond to concerns about Zika to women travellers.

REFERENCES

Australian Department of Foreign Affairs and Trade (DFAT) (2020) *Consular State of Play 2018–2019*. DFAT: Canberra.

Avraham, E. and Ketter, E. (2008) *Media Strategies for Marketing Places in Crisis*. Elsevier: Amsterdam. pp. 19–51.

Avraham, E. and Ketter, E. (2016) *Tourism Marketing for Developing Countries*. Palgrave Macmillan: Basingstoke, UK. pp. 109–133.

Beirman, D. (2009) 'Crisis and post-crisis destination marketing strategies', in Pforr, C. and Hosie, P. (eds), *Crisis Management in the Tourism Industry. Beating the Odds*. Ashgate: Farnham, UK. pp. 207–224.

Beirman, D., Upadhayaya, P., Pradhananga, P. and Darcy, S. (2018) 'Nepal earthquake in the aftermath of the April/May 2015 earthquake and aftershocks. Repercussion, recovery and the rise of new tourism sectors', *Tourism Recreation Research*, 43(4): 1–11.

Buhalis, D. and Darcy, S. (2010) *Accessible Tourism Concepts and Issues*. Channel View Publications: Bristol, UK.

Centers for Disease Control and Prevention (CDC) (US Government) (2020) www.cdc.gov. Accessed 13 December 2020.

CNBC (2009) Flu Concern for US Business. https://www.cnbc.com/2009/09/30/h1n1-vaccine-is-top-flu-concern-for-us-business.html. Accessed 18 May 2021.

CNN (2009) Top health Stories for 2009. http://www.cnn.com/2009/HEALTH/12/30/top/health.stories/index.html. Accessed 14 May 2021.

D'Olivera, W.K., de Franca, G.V.A., Carmo, E.H., Duncan, B.B., de Souza Kuchenbecker, R. and Schmidt, M. (2017) 'Infection-related microcephaly after the 2015 and 2016 Zika virus outbreaks in Brazil: a surveillance based analysis', *The Lancet*, 390 (10097): 861–70.

Flynn, C. (2020) *Presentation to World Tourism, Culture and Heritage Association*. Keeping the Dream Alive series presentation. www.wtach.org. Accessed 9 December 2020.

Henderson, J.C. (2003) 'Managing a health-related crisis. SARS in Singapore', *Journal of Vacation Marketing*, 10(1): 67–78.

Hospitality Net (2003) *Pacific Asia Travel Association Launches Project Phoenix*. 10 June 2003.

International SOS (2020) https://internationalsos.com. Accessed 9 December 2020. International SOS is one of the world's largest tourism safety companies.

Israel, B. (2017) Seven months after Rio Olympics Zika continues to plague babies in urban slums. https://www.sciencedaily.com/releases/2017/03/170324104927.htm. Accessed 18 May 2021.

Lewnard. J., Consuales, G. and Ko, A. (2016) 'Low risk of Zika spread due to 2016 Olympics in Brazil', *Annals of Internal Medicine*. https://doi.org/10.7326/M16–1628. Accessed 9 December 2020.

New York Timea (2009) US declares public health Emergecy over swine flu. http://www.nytimes.com/2009/04/27/world/27flu.html. Accessed 18 May 2021.

Nielsen, C. (2001) *Tourism and the Media*. Hospitality Press: Melbourne.

Roos, R. (2011) *Study Puts Global 2009 H1N1 Infection Rate at 11%–21%*. University of Minnesota Center for Disease Research and Hospitality, 8 August 2011.

Safety of Life at Sea (SOLAS) (2020) *Convention of the International Maritime Organisation*. http://imo.org/en/About/Conventions/Pages/International-Convention-for-the-Safety-of-Life-at-Sea(SOLAS)-1974.aspx. Accessed 2 December 2020.

Smith, J. (2013) *Will Accessible Tourism Soon be the World's Largest Travel Market?* World Travel Market Responsible Tourism. http:/blog.wtmresponsibletourism.com/2013/09/17accessible-tourism-largest-travel-market. No longer online.

UN World Tourism Organisation. (2020) *Accessibility*. https://www.unwto.org/accessibility. Accessed 12 May 2021.

Wilder-Smith, A. (2006) 'Tourism and SARS', in Wilks, J. and Pendergast, D.P. (eds), *Tourism in Turbulent Times. Towards Safe Experiences for Visitors*. Elsevier: Amsterdam. pp. 53–62.

World Health Organisation (WHO) (2004) *Report on SARS*. https://www.who.int/csr/resources/publications/WHO_CDS_CSR_ARO_2004_1.pdf Accessed 11 May 2021.

World Health Organisation (WHO) 2016 The Spread of Zika 2013–2016. https://who.int/bulletin/online_first/16-171082/en. Accessed 18 May 2021.

World Health Organisation (WHO) (2019) Zika Epidemiology Update July 2019. https://www.who.int/emergencies/diseases/zika/zika-epidemiology-update-july-2019.pdf. Accessed May 18 2021.

World Health Organisation (WHO) (2020) https://who.int. Accessed 9 December 2020. This site is a portal to a large amount of material on global health and pandemics issues.

Yad Sarah Israel (2020) https://yad-sarah.net/?lang=eng. Accessed 9 December 2020.

Yates, M. (2006) 'Project Phoenix: A benchmark for reputation management in travel and tourism', in Wilks, J., Pendergast, D. and Leggat, P. (eds), *Tourism in Turbulent Times: Towards Safe Experiences for Visitors*. Elsevier: Amsterdam. pp. 263–74.

8
THEME 5: CRIME AND TOURISM

INTRODUCTION

The real and perceived risk of exposure to crime and its prevalence is an influential factor in travellers' assessments of the overall safety of a tourist destination and destination choice (Beirman, 2003; Tarlow, 2014b). In some destinations, highly publicised criminal acts targeting tourists have led to sudden declines in visitation to that destination. This chapter will include two case studies, the first on Tourism Oriented Police Services and a case study focusing on the impact of Brazil's crime problems during the 2016 Rio Olympic Games.

As discussed in Chapter 5, there is a significant distinction between terrorism and crime but the presence or perception of either is a deterrent for tourists. While both terrorism and crime inflict loss through the deprivation of their victims' life, physical security, wealth, reputation or property, the motivations are significantly different. As discussed in Chapter 5, terrorists generally commit their acts openly, in the name of or in support of a political, ideological, social or religious cause. The commission of a terrorist act usually involves the perpetrators or their allies, publicising and promoting themselves and their cause. Some terrorists are prepared to accept retribution or trial for their actions as a means to advocate their cause. A sub-category of terrorists is even prepared to die in the commission of their act of terrorism (suicide bombers) as a form of martyrdom for their cause or faith. Some Islamist groups see the concept of the *shaheed*, a martyr in the name of *jihad* who sacrifices his or her life while killing 'enemies of Islam', as an action or an individual perpetrator to be revered.

By contrast, criminals seek to kill, hurt or deprive their victims anonymously, for personal or group/gang gain and to avoid identification as a perpetrator. Following the commission of the crime, they hope to avoid legal or financial accountability or retribution. They certainly want to live to enjoy the 'fruits' of their action. The crime rate of a particular country or jurisdiction is measurable through police and legal records. However, while official crime statistics provide guidance on crime rates, they are rarely a complete picture of crime levels in any given jurisdiction. Most crimes remain unreported and unrecorded for a range of reasons. These include embarrassment or perceived social stigma on the part of victims, through to a lack of confidence that police and judicial authority will pursue and punish criminals. There is a link between political stability and crime rates. The extent to which corruption exists in a society may also influence the conduct and effectiveness of law enforcement. As this chapter demonstrates, tourists are highly vulnerable to crime, especially opportunistic crime (Lisowska, 2017; Pizam, 1999).

Other factors that come into play include:

1. Prevailing social, religious and cultural norms in a society.
2. The quality, reliability and deterrence of law enforcement and the legal system.
3. The state of the economy, distribution of wealth and rate of employment.
4. The extent to which official corruption is endemic in a society.
5. The extent of limitations against alcohol, narcotics or dangerous drugs.
6. The extent of citizens' access to deadly weapons.
7. Social, racial, religious, political and ethnic tensions.
8. Significant and entrenched social and economic inequalities within the host society.
9. The prevailing political system and its approach to law and order.
10. Significant educational inequalities or inequalities of access to education.
11. Sexual mores that tolerate crime against women and people who are homosexual.

There are many more social and economic factors associated with criminality in a given country but the above points provide indicators, which could help to assess the likelihood of crime within a given society. A tourist's lack of knowledge of the culture of a destination community can result in them being an easy target for scams. Lonely Planet (2020) has outlined a range of common scams, which tourists fall for. A popular scam in China is an invitation to a 'tea festival' in which the tourist is invited to a tea house to sample a variety of teas and then charged an exorbitant price for the tasting. This is a minor scam, compared to some scams in which fraudsters and criminals deprive tourists of their money, credit cards, passports, goods, and in extreme cases, endanger their lives.

From an academic perspective, Professor Abraham Pizam and Professor Yoel Mansfeld are widely regarded as the pioneers of research on the linkage between tourism and crime. They edited one of the earliest and still one of the best books as a forum for a range of scholars, discussing the interrelationship between crime and tourism (Pizam and Mansfeld, 1996).

GLOBAL CRIME RANKINGS

Crime statistics are far from reliable, due to the fact that many crimes are unreported and, in some jurisdictions, policing and legal processes are minimal. Figure 8.1 is therefore a guide rather than a definitive indication of crime. It is noteworthy that many countries with high crime rates continue to attract tourists. These include South Africa, Brazil, Trinidad and Tobago and Jamaica. However, in the case of Venezuela, with the world's highest crime rate, the country's political, economic and social instability is a major contributing factor to its demise from a formerly oil rich and wealthy country, once attractive to tourists, to a failed state. Afghanistan and Syria, which also feature in the global top ten in relation to high crime rates, are in the midst of civil wars.

As discussed elsewhere in this chapter, some countries with acknowledged high crime rates do advise visitors to avoid those areas within the country which are crime hotspots. The fact that a country may have a high national crime rate does not necessarily mean the entire country is infested with crime. Most experienced travellers know that some parts of a country are virtually crime-free, while others are known as hotspots due to the factors outlined in the introduction to this chapter.

Figure 8.1, which is sourced from World Population Review (2020), is based on reported crime covering all key crime categories, from murder and assault through to theft. It's based on crimes as reported to police in the various countries. The figure includes the crime rates per 100,000 people for the top ten countries, and others with their ranking are selected. Readers interested into delving into crime statistics in more detail would find that police and crime statistics in individual countries and Interpol will provide more specificity and focus on specific categories of criminality.

Ranking	Country	Crime rate per 100,000 people	Population (millions)
1	Venezuela	84.36	28.536
2	Papua New Guinea	80.04	8.947
3	South Africa	77.29	59.308
4	Afghanistan	76.97	38.928
5	Honduras	76.65	9.904
6	Trinidad and Tobago	72.43	1.399
7	Brazil	68.31	212.560
8	Guyana	68.15	0.786
9	El Salvador	67.84	6.486
10	Syria	67.42	17.500
11	Jamaica	67.2	2.961
13	Angola	65.75	32.166

(Continued)

Figure 8.1 (Continued)

Ranking	Country	Crime rate per 100,000 people	Population (millions)
18	Bahamas	62.74	0.393
20	Kenya	61.73	53.771
26	Malaysia	58.55	32.365
35	Maldives	54.61	0.544
37	Mexico	53.31	128.932
50	United States	47.7	331.002
52	France	47.37	65.273
54	Egypt	46.65	102.334
58	Indonesia	46.06	273.523
64	United Kingdom	44.54	67.886
65	India	44.42	1,380.005
66	Italy	44.24	60.461
72	New Zealand	42.19	4.822
74	Australia	41.67	25.499
78	Canada	40.64	37.742
79	Russia	40.6	145.934
80	Sri Lanka	40.23	21.413
81	Thailand	40.01	69.799
83	Turkey	39.5	84.339
90	Germany	35.14	83.783
96	Spain	32.33	46.755
100	Israel	30.44	8.655
121	Japan	21.67	126.477
125	Hong Kong	20.91	7.496
128	United Arab Emirates	15.45	9.890
129	Taiwan	15.28	23.816
130	Qatar	11.9	2.881

Figure 8.1 Global crime rates 2020 – top 10, plus selected countries

Source: World Population Review, 2020. NB: No figures available for China and many other countries. Figure developed by the author

ANOMIE AND TOURISTS' VULNERABILITY TO CRIME

Dr Peter Tarlow has devoted much of his 40-year-plus research career to the issue of tourism safety and the interrelationship between crime and tourism. A core issue in this chapter is the vulnerability of tourists to crime especially when they travel internationally.

Tarlow uses the term 'anomie' to describe tourist vulnerability to a range of crimes (Tarlow, 2014b).

Anomie involves a sense of unfamiliarity or unease in an environment in which the individual is inherently unfamiliar. Most international travellers can readily identify with anomie they experience when they arrive in a new country for the first time, even if they had researched it prior to arrival. The language, dress, signage, currency, aromas, customs arrival procedures and culture are all unfamiliar. Independent tourists who chose not to pre-arrange an arrival meeting, or accommodation package starting at a gateway point, may be approached by many people offering accommodation, transport and a range of other travel services. Some of these are legitimate operators but a portion are touts who prey on naïve tourists by claiming to offer accommodation and transfers of a standard and value well below the promised product.

Arriving at a new destination with no pre-booked accommodation arrangements can be a daunting experience at the best of times. Newly arrived tourists afflicted by jetlag, exhaustion, confusion arising from unfamiliarity with a new and unfamiliar destination, culture and language experience anomie. Many tourists, often unintentionally, display conspicuous evidence of wealth on their person, through the clothes they wear, jewellery, watches, money pouches worn outside their clothing, expensive spectacles, iPad and expensive pens. They are readily identifiable targets of great opportunity for petty criminals engaged in pickpocketing, assault, robbery, fraud and scams. Even those tourists who believe they are blending in to local dress tend to stand out when arriving in an unfamiliar destination. Tourists are widely regarded as easy or soft targets for criminals.

The type of crime committed against tourists encompasses the full range of criminality from serious crimes involving murder, kidnapping/abduction, rape, sexual assault and harassment, assault causing grievous bodily harm, carjacking and extortion to opportunistic crimes of robbery, pickpocketing, fraud, identity theft, scamming of credit and debit cards and overcharging by tourism and hospitality businesses. Some tourists are persuaded by local criminals to engage in conduct, which they may be unaware is illegal in a destination community, including child prostitution, the consumption of alcohol and drugs, proscribed forms of gambling activity, cockfights and driving vehicles without a legitimate licence. Tourists are sometimes unwitting accomplices for the smuggling of drugs, weapons and contraband.

In some tourist destinations, crime against tourists has a negative impact on the reputation of a destination. Tourism destinations in which frequent acts of violence against tourists occur, especially murder, rape, assault and carjacking, can lead to negative media coverage of that destination. Many Caribbean countries, popular among tourists from the USA, Canada and Europe, also experience high crime rates (most of this within their own societies). However, when instances of crime target tourists they tend to be widely reported in the media of the victim's home country (Avraham and Ketter, 2008).

Tourists have access to a number of sources of information to prepare them for a range of risks, including crime prior to arrival at the destination/s of their choice. Most government travel advisories include a section that outlines the main crime risks applicable to a given destination. This section usually warns tourists about behaviour with significant legal sanctions, which may be different from their country of citizenship. These include strict regulations forbidding alcohol consumption or inappropriate dress (especially applying to women) and overt displays of affection in some Muslim countries, which could be subject to arrest and imprisonment. As discussed elsewhere, legal penalties against drug use or drug importation are far more severe in many Asian countries than they may be in Europe, North America or Australasia.

In addition to government travel advisories, tourism guide books and online resources with destination information usually include a section on crime and scam risks, which apply to destinations. Some countries, including South Africa, Mexico, Vietnam and Trinidad and Tobago, issue pre-arrival information warning tourists about the most likely crime threats and suggest measures they can take to minimise exposure to them.

RISK MANAGEMENT AND TOURISM CRIME

One form of risk management available to tourists is travel insurance, which can cover lost or stolen valuables. In the event of theft, it is important for travellers to have evidence of a report of theft to the police. While there is minimal likelihood of stolen goods being retrieved, insurers will provide financial compensation when documentary evidence is provided. Careful use of credit and debit cards and advice to credit card providers when a card is stolen can prevent loss or annul payments made on stolen credit cards. The use of cash in many industrialised destinations is becoming increasingly uncommon, although there are many developing countries which only accept cash transactions. In the event of lost and stolen passports, diplomatic legations or their authorised representatives can assist in replacing passports at a cost. The travel advisories of most tourism generating countries urge their citizens to purchase travel insurance. They also provide advice about the main crime risks, which apply to any given destination and some preventative measures travellers can take to minimise their exposure to crime.

Crime is not only committed against tourists but is sometimes perpetrated by tourists in some destinations. Certain forms of tourism generate an environment conducive to criminality. A common crime committed by many international tourists involves the smuggling of contraband or prohibited products at the point of entry. Popular TV series such as 'Border Security' (which features international travellers arriving in a range of destinations by land, sea and air) highlight the tendency of some tourists to bring in their favourite local foods (even when bringing in food is explicitly banned) based on a concern that these culinary favourites may not be available in their chosen destination. Some tourists bring in amounts of cash, cigarettes, gift items and liquor above the permissible duty-free limits. Island nations such as Australia, New Zealand and the UK take special care to ensure that plant, animal and insect material does not enter and endanger the national environment. Attempts by tourists to smuggle in most of the above illegal products are subject to heavy fines, duty payments, confiscation and in extreme cases, deportation and imprisonment.

Far more serious criminality involves attempts by tourists to smuggle, sell or use illegal narcotic drugs. Some of these incidents have resulted in perpetrators being subject to lengthy terms of imprisonment and on rare occasions, judicial execution. In April 2005, a group of nine Australian citizen who travelled to Bali (Indonesia) were arrested by customs officers at Denpasar airport. They had attempted to smuggle 8.3 kg of heroin into Bali. All were sentenced either to death or to lengthy prison terms. After a very lengthy and highly publicised legal process, the two ringleaders, Andrew Chan and Myuran Sukumara, were executed by firing squad on 29 April 2015. In Australia, there were some concerns among Bali specialist tour operators that the executions may lead to a slump in Australian tourism to Bali. This did not occur but the long-drawn-out saga of the 'Bali Nine' proved a salutary lesson to many young Australians who were under the impression that Bali was an anything-goes party destination. As discussed in this chapter, Bali's police chief, Dr Petrus Golose, developed an extensive training programme for members of Bali's police force to sensitise them to the challenges of policing tourists and foreign tour operators to respect the sensitivities of Balinese society (APEC, 2017).

Incidents of this nature have occurred in ASEAN countries with zero tolerance for the importation and use of hard drugs such as heroin, cocaine or methamphetamine. A small number of tourists, including some from Western countries, involved in drug smuggling were executed in recent years in Singapore, Malaysia, Thailand, Cambodia and Indonesia. Many thousands more tourists have been imprisoned for drug-related crimes in Asia and elsewhere.

Another common crime, associated with and sometimes perpetrated by international visitors, involve crimes related to sexual activity, including paedophilia, the trafficking of prostitutes and the distribution of pornography. Gambling and gambling-related crime is another field of criminality engaged in by a minority of tourists. Some criminals, seeking to escape prosecution for criminal activity in their countries of citizenship, seek refuge in other countries. Until the overthrow of former Filipino President, Ferdinand Marcos in 1986, sleaze tourism centred on gambling and prostitution had a high profile amongst a subculture of visitors to the Philippines who were attracted by the prospect of 'thriller in Manila' holidays focusing on sex, paedophilia and gambling. Sex tourism, associated with crime syndicates (some local and some foreign), remains a problem in a number of developing countries and the hiring of under-aged female and male prostitutes conflicts with the traditional values of some of these countries (Beirman, 2003).

CRIME – ITS IMPACT ON TOURISM AND DESTINATION REPUTATION MANAGEMENT STRATEGIES

There are many countries in the world with serious crime issues. Ironically, some of these countries remain very popular tourism destinations. A strategy developed by South African Tourism in the late 1990s acknowledged the high rate and severity of crime in that country but pointed out that most serious crime occurred in specific districts. In 1998, South Africa had the highest rate of serious crime in the world, applying to murder, attempted murder, rape and assault causing grievous bodily harm and carjacking. By 2020 it had the world's third highest crime rate after Venezuela and Papua New Guinea.

Most crime in South Africa was a consequence of the massive social inequalities in the country resulting from years of Apartheid (1948–1994). While some of the criminality involved poor people targeting the affluent, the squalor and poverty common in so many shanty African townships was a breeding ground for 'Black on Black' crime. Not even Africa's much esteemed first post-Apartheid government, led by President Nelson Mandela, could rapidly reverse decades of extreme social and economic inequality and re-house millions of residents of shanty towns into better planned and higher standard residential neighbourhoods. While South Africa's crime rate is statistically high, most violent crime tends to be concentrated in specific districts and localities. Avoidance of these districts and localities is the key to minimising exposure to crime in South Africa.

South African Tourism Board (SATOUR) provided advice to tourism professionals and tourists alike on strategies for tourists to minimise their exposure to crime hotspots and unwelcome attention from criminals (Beirman, 2003). The strategy was highly successful in reassuring tourists that if they followed the advice of South African tourism they could enjoy a safe trip in South Africa. From the late 1990s and until the Global Financial Crisis negatively impacted on tourism to South Africa in 2008, the number of international tourists to South Africa grew steadily each year and rose from 5.8 million in 1998 to 9 million in 2007.

In 2006, the unwanted title of the world's highest crime rate passed to Trinidad and Tobago. Due to a number of assaults directed at tourists from the USA (Trinidad and Tobago's main source market) the tourism ministry of Trinidad and Tobago was concerned that tourists would abandon the country. In common with most Caribbean island countries, Trinidad and Tobago have a high level of economic dependency on tourism. Tobago, the small island to the north of Trinidad, had a 97 per cent economic dependency on tourism while Trinidad's economy was 15 per cent reliant on tourism. Trinidad and Tobago's Ministry of Tourism was so concerned about the negative impact of crime on the destination's image and reputation that it sought external assistance. In May 2006, Trinidad and Tobago Ministry of Tourism invited leading UK management consultant and crisis management specialist, Dr David Perl, to run a series of workshops for tourism professionals. Dr Perl, a qualified medical practitioner, who established the UK-based risk and crisis management consultancy Docleaf, in which he was the Principal, developed a strategic programme for the Ministry and the wider tourism industry to counter the negative impact of the crime spike on Trinidad and Tobago's vital tourism industry. Part of the project involved training sessions of two days each for all tourism professionals in Trinidad and Tobago funded by the tourism ministry. The Ministry took the project so seriously that Tourism Minister, Mr Howard Chin Lee, personally participated in the full two-day workshop in the capital city, Port of Spain. Dr Perl, in turn, asked the author to assist him in presenting the workshops.

The project involved applying and adapting the strategies successfully developed by South Africa to a Trinidad and Tobago context. There were some key differences in that some of the criminality in Trinidad and Tobago targeted tourists on resort property. Consequently, in addition to the public relations dimension, the strategy focused on building a more positive image of Trinidad and Tobago to its key international source markets. There was significant emphasis on upgrading hotel and resort security in which the national police worked closely with resort owners and resort security.

A key outcome of the training sessions was that all sectors of the tourism industry emphasised safety and making tourists aware of how they could make the most of their visit to Trinidad and Tobago, while minimising their vulnerability and exposure to crime. In fact, all government and private sector tourism promotional material about Trinidad and Tobago between 2006 and 2010 included a section on safety and security tips to travellers, written and presented in a relaxed Caribbean fashion.

Caribbean island states have a very high dependence on tourism by global standards. Of the 25 countries with the world's highest level of economic dependence on tourism (based on tourism's portion of GDP), 11 are Caribbean countries (Frost, 2019). Many Caribbean countries are also notable for high crime rates (World Population Review, 2020) with five out of the world's top 25 countries based on high crime rates located in the region. Trinidad and Tobago is ranked at 6th and Jamaica 11th. As both of these Caribbean island countries are heavily reliant on tourism for their national economies, this is not a position either government would welcome. The Caribbean Tourism Organization, which represents 24 Caribbean island countries, has placed tourism recovery high on their agenda and is developing a united strategy to combat crime as a high priority.

Case Study 8.1

Tourism Oriented Police Protection and Security (TOPPS) – A Global Policing Challenge

There are many challenges for police in relation to tourism. Police not only have to protect tourists from those locals who may prey on them as soft targets for opportunistic crime but they are often expected to protect tourists from their own errors of judgement and sometimes the stupidity and cultural insensitivity of their behaviour.

Police forces all over the world have a prime duty to protect the citizens within their jurisdiction from crimes against their person and property. This responsibility also extends the protection of non-resident people who visit their jurisdiction. While some tourists make the effort to research the destination they visit and acquire an awareness of the laws, customs and behavioural norms that apply to the destination they visit, most do not. Police and security officials need to understand some of the sensitivities and vulnerabilities of tourists who may lack awareness, knowledge and sensitivity towards the legal, social and transactional norms of a society in which they are a stranger and generally don't understand the local language.

Figure 8.2 Dr Peter Tarlow

Source: Peter Tarlow, 2014a. Used with permission.

Dr Peter Tarlow is one of a number of extraordinary and talented people drawn from other fields to the tourism industry. Since 1990, he has been teaching courses on tourism and crime to police forces and security professionals in many countries. His path to this work is unusual. Tarlow is an ordained Rabbi based in College Station, Texas, US, affiliated with Texas A & M University. His fluency in Spanish and Portuguese has enabled him to present his tourism security programmes to police and security personnel in almost every country in North and South America and the Caribbean – in their own language.

Tarlow has built a network of security and tourism experts under the umbrella of his consultancy, Tourism and More (2020), which is able to contribute tourism-oriented security expertise to assist civilian police, event security personnel, counter-terrorism units and specialised military units. Since 1992, he has been running the annual International Tourism Safety Conference in Las Vegas but has spoken at countless conferences with a focus on tourism security all over the world. The author was a fellow speaker with Tarlow at such conferences in Malaysia, Canada and Jamaica. Tarlow's primary role is the running of training workshops with police forces and security professionals all over the world, primarily in the Americas and the Caribbean. A highlight of his career was working closely with the Brazilian police and security forces to prepare tourism oriented security measures for the 2014 World Cup and the 2016 Rio Olympics discussed in Case Study 8.2 of this chapter.

In the context of this chapter, Tarlow's major contribution to tourism policing and security is the programme he has been developing since 1990 known as TOPPS (Tourism Oriented Policing and Protection Services). Tarlow has written widely on TOPPS (Tarlow, 2014a), and his Tourism and More website (2020) and regular e-publication *Tourism Titbits* contain a vast amount of information about it.

The essence of TOPPS involves explaining to police and security personnel the many vulnerabilities of tourists who visit their jurisdiction, and to assist them in developing proactive, real-time and response strategies to minimise the likelihood of crime against tourists, treat the risk of crime and respond to victims of crime. TOPPS also explains to police and security people that tourism is a major economic asset for their jurisdiction and that effective policing, which protects tourism, benefits the local community.

These vulnerabilities include:

1. Ignorance of local customs and social mores.
2. Lack of local language skills.
3. Lack of knowledge of prevailing crimes, high crime areas and scams, which are common knowledge to locals in the destination.
4. Tourists and their valuables are often far more conspicuous than tourists themselves think. Hand-held smart phones, tablets, bumbags, shoulder bags, backpacks and handbags containing valuables worn outside clothing, expensive clothing, footwear, jewellery and watches are tempting targets for opportunistic theft from pickpockets and bag snatchers.

(Continued)

5. Tourists referring to maps and generally appearing confused are a tempting target for scammers, fraudsters and touts.
6. Lack of knowledge of laws prevailing in a destination. This sometimes results in tourists being arrested or punished for dress and actions that are permissible in their home countries.
7. Lack of knowledge of how much goods and services cost in a destination. Tourists are often easy marks for unscrupulous traders and retailers, eateries, taxi drivers, people who pose as 'tour guides' and touts. The economic naïvety of tourists makes them a soft target for the charging of extortionate prices for many goods and services. Many tourists fail to understand that price bargaining is a common practice in many destinations and while it's not necessarily a crime to fleece the unwary, it does little to enhance the tourist experience.

TOPPS programmes explain how police and security officials can most effectively work with tour operators, tour guides, destination managers and hoteliers to assist and advise tourists on measures taken to minimise tourist exposure to crime. Most hotels have a room safe or a hotel safe to store valuables so that conspicuous displays of wealth are minimised. Tour operators, destination marketing authorities and hotels can provide advice on local crime risks in the language of the visitors. TOPPS has a strong focus on collaborative approaches between police and the tourism industry in host societies to minimise tourism vulnerability to crime.

TOPPS has a strong emphasis on event security and identifying those areas within a jurisdiction where the likelihood of crime against or committed by tourists was greatest. TOPPS essentially guides police to prioritise crime risks involving tourists and to work with tourism infrastructure and businesses to enhance security. Increasingly, TOPPS training is conducted with counter-terrorism police units, especially in preparation for mega events focusing on event security and protecting the influx of tourists, which often accompany major events.

In most cities in the world, there is a close working relationship between the counter-terrorism police units, local police and event organisers in relationship to major cultural, sporting and commemorative events. Wayne Middleton, who has been appointed as the risk manager for a wide range of events throughout the world, including the 2000 Sydney Olympics and as a consultant to the 2012 London Olympic Games, has frequently pointed out that collaboration in security for these events involves police forces, military and event security specialists (Middleton, 2019). Middleton's company was among the first in the world to introduce an online programme known as Risk Sense (Reliance Risk, 2020) for event risk and security management.

In many countries with high rates of tourism, affiliate arms of police forces are units known as tourism police. Tourism police operate in many countries globally. Although very few tourism police have the power to arrest, they are a visible source of advice and assistance to tourists and work closely with local police.

At the very least, their visible presence at airports, railway stations, bus stations and popular tourism attractions acts as a deterrent to crime against tourists. Many are multilingual and in Thailand especially, which until 2019 was hosting 40 million international tourists, the Thai government recruited many non-Thais who could speak the languages and understand the cultures of foreign tourists as tourism police. A visible tourism police presence in popular tourism attractions is balanced by covert security measures in the form of technologically-based surveillance systems and plain-clothed police.

Peter Tarlow's TOPPS programme was adopted by local, state and national police forces in every country in North and South America and the Caribbean. It has also been widely adopted in the UK, many European countries, Kenya and other African countries. Some police forces in India are also adopting it.

In Bali, Indonesia, Bali's police chief, Dr Petrus Golose (now the head of Community Policing in Bali), developed a community and tourism-oriented police training system that shares many common characteristics with TOPPS. His brilliant presentation at the APEC Counter Terrorism and Tourism conference held in Bali in May 2017, during which he explained his programme, was a hard act to follow for the author, whose presentation followed Bali's Police Chief (APEC, 2017). Dr Golose has a PhD in criminology and specialises in community policing, although until that conference, he had been unaware of either TOPPS or Tarlow. The enhancement of tourism-oriented policing in Bali became a major priority after the mass casualty terrorist attacks in Bali of 2002 and 2005, discussed in Chapter 5.

As Bali's economy is heavily dependent on tourism, Dr Golose, in common with TOPPS, sought to develop a policing programme involving a high level of collaboration between the police with the local tourism industry and local community organisations. Bali is Indonesia's most popular destination for international visitors. Although most tourists are respectful of Bali's distinct Indian influenced culture, there is a significant minority of tourists (many of them young Australians) who have treated a visit to Bali as a stay in a tropical pub and behave accordingly. The bad behaviour of a minority of tourists is reflected by the growth of local crime, much of it directed at tourists. Dr Golose believes that if local people and tourism operators are made more aware of cultural mores and clearly communicate the laws which apply to Bali, this will reduce criminal and insensitive behaviour by tourists and Balinese within the host society. He has established very good relations with all community groups in Bali and with the Indonesian government. His ability as a singer and entertainer has worked as an unusually effective icebreaker in his dealings with a wide range of stakeholders. His formal presentation at the 2017 APEC Counter Terrorism and Tourism Conference concluded with a rendition of Louis Armstrong's hit song, 'What a Wonderful World'. He sang it very well!

In essence, TOPPS and Dr Golose's community and tourism policing programmes both incorporate many common-sense elements in countering crime committed

(Continued)

against and by tourists. Most significantly, both programmes stress the importance of policing in collaboration with a broad spread of community groups and with local and international tourism companies that service the destination. This collaboration enables police and security forces to ascertain and prioritise the main crime threat for tourists.

Figure 8.3 Dr Petrus Golose, Police Chief, Bali, Indonesia

Source: Bali Police, 2017. With kind permission, Dr Petrus Golose

CYBERCRIME AND TOURISM

Cybercrime is a necessary inclusion in any chapter related to crime and tourism risk. Cybercrime is that criminal activity which seeks to damage, disrupt and steal data and money from legitimate businesses and consumers. Although it is arguable that cybercrime is not necessarily a cause of crises at the destination level, cybercrime has the capacity to inflict considerable reputational damage and operational disruption to tourism businesses, and cause distress and inconvenience to travellers and travel companies alike.

The hacking or theft of data and databases, scamming of financial information, money and financial transactions has become a very real threat to all businesses, including tourism businesses. In the relatively free online world, the spreading of disinformation has the potential to damage the reputation of destinations and tourism enterprises.

As the tourism industry has become increasingly reliant on online bookings, customer data and financial transactions, actions that disrupt and steal (or hackers who insert viruses into online data) present a real danger to businesses and consumers alike. The growing reliance on the part of hotels, airlines, cruise operators, tour operators, event organisers of online data, booking information and financial transactions makes them vulnerable to data theft and data disruption. On occasions when computer systems (especially for airlines) crash, the result is panic and confusion at airports worldwide. Data theft or disruption can impact on the status of bookings for a wide range of tourism and hospitality services.

The majority of the world's tourism transactions operate via global distribution systems (GDS) dominated by Amadeus, Sabre and Travelport/Galileo. These companies link the services of nearly all the world's airlines, hotels, cruise companies with online and traditional travel agencies and booking platforms in addition to selected corporate clients. They have links to a myriad of other travel-related services. In order to protect data, bookings and transactions, the GDS companies invest many millions of dollars in data security technology, back-up systems and encryption to minimise the risk of cyber-criminals attacking data (Amadeus Global Distribution System, 2020). However, there is no guarantee that any of these protective measures are invulnerable.

Related to cybercrime is document forgery. In recent years, most countries have moved to issue biometric passports in which data about the passport holder is included in a chip inserted into the passport. This has made the forging of passports exceedingly difficult. The Australian Passport Office, in common with all 'five eyes' countries have included over 100 security measures into each passport, some of which are explained in passport documents, most are not (Australian Passport Office, 2020).

Forgery of documents, paper/plastic currency, credit and debit cards is a common form of crime made increasingly difficult by technological counter measures. Some criminals and even a few state intelligence services employ specialist master forgers to undertake these acts of sophisticated forgery, which amount to theft and fraud on a grand scale. There is a constant but little publicised battle in the cyber world between increasing skilled hackers, cyber criminals and forgers and those who seek to prevent them.

Case Study 8.2

Crime and Tourism During the 2016 Rio Olympic Games

Brazil is one of the most varied and interesting destinations in the world but the number of international tourism arrivals has never reflected the range and quality of this vast country's attractions, which include the Amazon rainforest, most of the Amazon River, pulsating cities, the Iguazu Falls and rugged mountains. Rio de Janeiro, Brazil's scenic second largest city is world famous for its magnificent harbour, glamorous beaches, Copacabana, Ipanema and Sugarloaf Mountain topped by the vast statue of Christ the Redeemer. It was always the hope of the Brazilian government that the expensive infrastructure investment involved in hosting the world's two greatest mega events, the World Cup football finals in 2014 and the Olympic Games 2016 would reap a tourism bonanza for Brazil. Despite the fact that both events ran successfully and safely, Brazil's reputation as a tourism destination was severely compromised by the country's high crime rate, and in the case of the 2016 Olympics, concerns about the Zika outbreak, discussed in Chapter 7. As Figure 8.4 attests, Brazil's international tourism arrivals experienced relatively modest growth between 2014 and 2019. The significant growth in tourism to Brazil so eagerly hoped by the Brazilian government and tourism industry on the back of the 2014 World Cup and the 2016 Olympic Games never materialised, although in the years leading up to the 2014 World Cup there was substantial growth in the number of international tourism arrivals. During 2020, Brazil's appeal for tourists was fatally compromised due to the misfortune of having one of the world's highest rates of Covid-19 cases. By 10 November 2020, there were over 5.8 million reported cases in Brazil (ranking it 3rd in the world for case numbers) and 162,000 deaths (ranking it 2nd in the world for number of deaths).

This case focuses on the impact of crime on tourism to Brazil during 2016, the year of the Rio Olympics. Earlier in this chapter it was revealed that Brazil's crime rate is the 7th highest in the world. Brazil, South America's largest and only Portuguese speaking country, has a rapidly growing population in excess of 212 million people. Brazil is a country characterised by massive social and economic inequalities. Although it is a country with great wealth and a substantial middle class, there are millions of Brazilians living in the largest cities of Sao Paulo (population 22 million), Rio de Janeiro (population 13.5 million) and the capital city Brazilia (population three million) who live in abject poverty. Many of these people reside in shanty towns known as favelas; many of them, especially in Rio, precariously located on the steep and flood prone slopes surrounding the main cities.

The massive and extremely visible gap between rich and poor, coupled with little government support for the countries aged, unemployed and low-income earners, have been the primary root causes of crime in Brazil's teeming cities. Most urban crime in Brazil tends to be opportunistic theft and assault, although the country has a high rate of murder, abductions, carjacking and severe assault and rape.

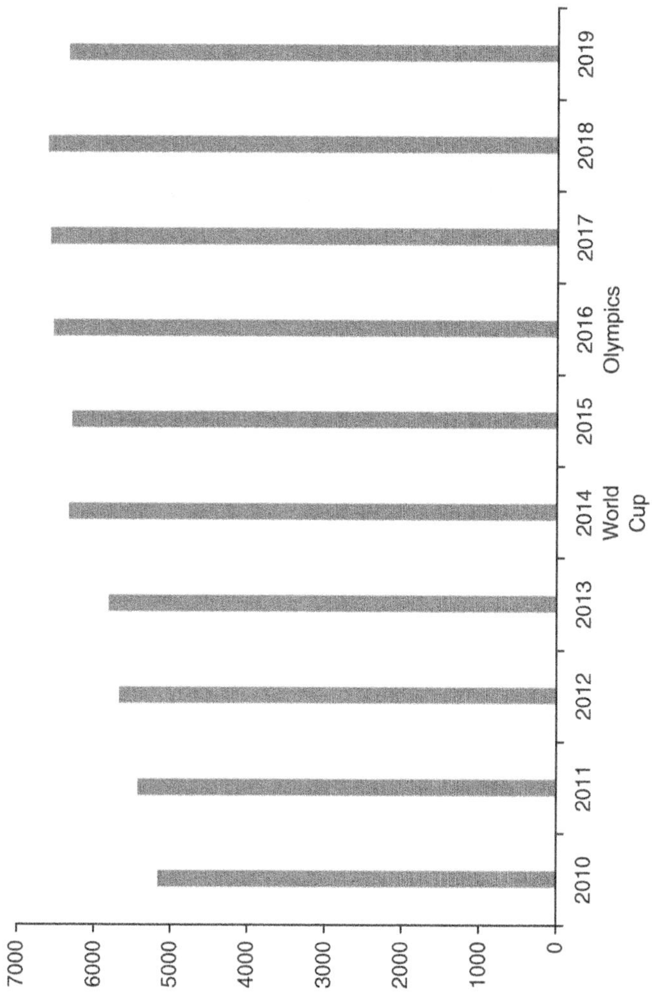

Figure 8.4 International tourism arrivals in Brazil 2010–2019 (thousands)

Source: UN World Tourism Organisation, 2020. Graph by the author

(Continued)

Preparations for the security of the 2016 Olympic Games events were managed with a high level of professionalism. Brazilian security services were well trained, including the TOPPS programme, and were supported by security experts from the USA, UK, Europe and Israel. However, the influx of athletes, their supporters and the media from all over the world shone a harsh light on social and crime problems in Rio specifically and Brazil more broadly. In the weeks leading up to and during the games themselves, the global media extensively reported incidents of street crime. In fact, some reporters and TV crews experienced acts of opportunistic criminality first hand when they were robbed of cameras and sound equipment, money and valuables (BBC, 2016; CNN, 2015; New York Times, 2016). Reports of visitors to Rio being robbed or assaulted became instant headlines in the countries whose citizens were among the victims. Although athletes were largely safe within the Olympic Village, those who ventured beyond the secure bubble of the Olympic Games occasionally became victims of crime.

In the hiatus period before the Games, when journalists were looking for a story, the media extensively reported on the clearly evident social inequalities in Brazil and the anti-government protests against the Games from the more deprived sections of Brazil's population. The appalling living conditions in Rio's favelas (slums) were cited as a prime motivator for street crime and many media reporters contrasted the lavish lifestyles of Rio's rich with the deprived lifestyles of the poor.

Academic research on the impact of crime on tourism during the 2016 Olympic Games indicated that there was greater concern about the risk of crime than terrorism or Zika (a concern mainly confined to women of child bearing age). Although there was a high degree of confidence that security for the Olympic Games and games events would ensure their safety, this confidence did not extend beyond the actual games themselves (Walters et al., 2017).

Not all media coverage was negative. Before, during and after the 2016 Olympic Games there was extensive positive media coverage, which highlighted the visual, cultural and historic attractions of Rio and Brazil. As most Olympic host cities do, certain events presented an opportunity to showcase the beauty of the city and its spectacular harbour.

Although international tourism to Brazil registered modest growth during and after the Rio Olympics between 2016 and 2018 it was well short of the high expectations of Brazil's government and tourism industry. Clearly, one of the risks of hosting a high-profile event is that the host city or country falls under a media microscope, which exposes its virtues, vices and social problems. Rio de Janeiro was certainly not the first Olympic host city that lacked a massive post-Olympics tourism boost. In fact, the majority of Olympic host cities derive far less post games tourism benefit than they expect (Adair and Frawley, 2013).

The challenge for Brazil's tourism industry and leadership was that no amount of PR skill or accentuation of the many positive qualities of Rio the city or Brazil could fully overcome the negative perceptions of Brazil's crime problems as the dominant negative perception of the destination. To its credit, Brazil's tourism industry did acknowledge the issue and did advise arriving and prospective tourists of measures that would minimise tourist exposure to and loss from crime. These included a recommendation that tourists are safer to visit Brazil in organised groups run by reputable tour operators.

SUMMARY AND CONCLUSION

This chapter has sought to cover the significant impact that crime has on tourism and the destination decisions tourists make. Crime can occur anywhere and many tourists recognise that there is a risk of exposure to crime whether they stay at home or travel. Criminal acts against tourists can occur in the statistically safest destinations. It is also important to recognise that some tourists are perpetrators of crime and should be subject to preventative measures.

However, countries with high crime rates face serious challenges in overcoming negative perceptions among prospective travellers. South Africa, Trinidad and Tobago, Brazil, Jamaica and the Bahamas are among countries with very high crime rates and significant numbers of tourists. A measure all these countries have employed, with varying success, is to acknowledge the problem and advise prospective and arriving visitors of measures they can take to minimize exposure to crime. One of many key measures is the avoidance of crime hotspots. Most countries have areas with especially high crime rates but travellers can visit regions in the same country which are almost free of crime. A high crime rate is not necessarily a barrier to tourism provided that tourists and the tourism industry are fully briefed on managing that risk.

In this chapter, many forms of crime have been discussed. Theft and assault can be as debilitating to a tourist's experience as being a victim of cybercrime, which compromises privacy, bookings and money. One of the key measures to combat crime against and by tourists is appropriate training of police and security professionals. TOPPS and the work of Bali's police are excellent examples of policing with an orientation towards addressing the vulnerabilities of tourists and the lack of knowledge many have of local laws, customs and culture.

From a tourism industry perspective, tourism professionals in tourism generating countries and in destinations should ensure that tourists visiting any destination are briefed about their legal obligations as visitors and are aware of the key crime risks that apply to the destination they visit. Government travel advisories are already addressing this issue but it should be an integral part of the toolkit of any person–person or online travel consultant.

Exposure to crime is a risk that every tourist takes but effective preparation for and treatment of that risk will ensure that most tourists travel safely in most destinations. Tour operators who are knowledgeable about a destination and are security conscious can offer increased levels of tourist confidence in destinations in which crime is acknowledged as a problem. The case study on Brazil demonstrates that the successful hosting of the greatest events on Earth is no guarantee that tourists will flock to a destination if a high crime rate is perceived as a detraction to tourism.

Discussion Questions and Issues

1. If you were advising the Brazilian tourism industry, how would you recommend they respond to their destination marketing to the high crime rate, especially with reference to the 2016 Olympic Games?
2. Some countries have had some success in maintaining strong international tourism visitation despite a high crime rate. Using one of these countries as an example, what measures and marketing measures would you recommend?
3. Discuss some of the special skills and sensitivities police and security professionals would require to best protect tourists at an attraction of your choice.
4. If you were running a wholesale tour operator to a destination with a high crime rate, how would you incorporate this information into your promotional literature without deterring tourists from visiting? Choose any destination and any type of tour.

REFERENCES

Adair, D. and Frawley, S. (2013) *Managing the Olympics*. Palgrave Macmillan: Basingstoke, UK.

Amadeus Global Distribution System (2020) https://amadeus.com. Accessed 9 December 2020.

Asia-Pacific Economic Cooperation (APEC) (2017) *Strengthening Tourism Business Resilience Against the Impact of Terrorist Attack*. Report of the APEC Counter-Terrorism Working Group's Workshop on 9–10 May 2017, Bali, Indonesia. APEC Counter-Terrorism Working Group, APEC Secretariat: Singapore. p. 94.

Australian Passport Office (2020) www.passports.gov.au. Accessed 29 November 2020.

Avraham, E. and Ketter, E. (2008) *Media Strategies for Marketing Places in Crisis*. Elsevier: Amsterdam. pp.19–43.

BBC (2016) Brazil Crime Rate and the Olympics. https://www.bbc.com/news/world-latin-america-36950412. Accessed 15 May 2021.

Beirman, D. (2003) *Restoring Tourism Destinations in Crisis. A Strategic Marketing Approach*. Allen & Unwin: Crows Nest (Sydney), Australia.

CNN (2015) Rio 2016. City grapples with wave of crime on beaches as Olympics near. https://edition.cnn.com/2015/10/06/sport/rio-2016-olympics-security-robbery/index.html. Accessed 16 May 2021.

Frost, N. (2019) These are the countries most reliant on your tourism dollars. *Quartz*,ct 10, 2019. https://qz.com/1724042/the-countries-most-reliant-on-tourism-for-gdp. Accessed 10 May 2021.

Lisowska, A. (2007) *Crime in Tourism Destinations: A Research Review*. University of Wroclaw (Poland), Institute of Geography and Regional Development Socio-Economic Geography Institution. *Toryzm/Tourism, Sciendo*, 27(1): 31–39.

Lonely Planet (2020). 10 Common Travel Scams and How to Avoid Them., http://www.lonelyplanet.com/thorntree/forums/travel-bloggers-and-pathfinders/topics/10-common-travel-scams-and-how-to-avoid-it 2020 Lonely Planet Global Inc. Accessed 9 December.

Middleton, W. (2019) *Presentation on Risk Management to UTS Students*. September, 2019.

New York Times. (2016) Security force of 85,000 hits Rio, unsettling human rights activists. Kaiser, A. and Jacobs, A. 8 August 2016 https://www.nytimes.com/2016/08/08/world/americas/rio-olympics-crime.html. Accessed 16 May 2021.

Pizam, A. (1999) 'A comprehensive approach to classifying acts of crime and violence at tourism destinations', *Journal of Travel Research*, 38(1): 5–12.

Pizam, A. and Mansfeld, Y. (Eds) (1996) *Tourism Crime and International Security Issues*. John Wiley and Sons: New York.

Reliance Risk (2020) *Risk Sense Software*. https://reliancerisk.com/risksense101. Accessed 9 December 2020.

Sabre Global Distribution System (2020) www.sabre.com. Accessed 9 December 2020.

Tarlow, P. (2014a) 'Tourism oriented policing and the tourism industry', *International Journal of Event Management Research*, 8(1): 1–18.

Tarlow, P. (2014b) *Tourism Security Strategies for Effectively Managing Travel Risk and Safety*. Elsevier: Amsterdam. pp. 23–47.

Tourism Authority of Thailand (2020) *Link to Tourist Police*. https://tourismthailand.org/Plan-Your-Trip/Useful-Call. Accessed 9 December 2020.

Tourism and More (2020) *Tourism Safety and Security Consultancy*. www.tourismand-more.com. Accessed 9 December 2020.

Travelport GDS System (2020) www.travelport.com. Accessed 9 December 2020.

Walters, G., Shipway, R., Miles, L. and Aldrugini, M. (2017) 'Fandom and risk perception of Olympic tourists', *Annals of Tourism Research*, 66, Issue C: 210–12.

World Population Review (2020) *Global Crime Statistics*. https://worldpopulationreview.com/country-rankings/crime-rate-by-country. Accessed 9 December 2020.

9

THEME 6: ECONOMIC CRISES

INTRODUCTION

Economic sustainability is one of the three pillars of sustainable tourism. From a sustainability perspective, tourism should be economically beneficial to the destination community. This means that income derived from tourism spent in a destination should utilise locally sourced and produced goods and services, create a range of jobs for the local community and add value to the local community's economic infrastructure. The 17 elements of the UN's Sustainable Development Goals (UNSDGs) (UN, 2020) provide an aspirational set of social, economic and environmental goals for sustainable tourism. Most of the 17 UNSDGs are focused on achieving economic and socio-economic outcomes for destination communities, including:

1. The elimination of poverty
2. The elimination of hunger
3. Good health and wellbeing
4. Clean water and sanitation
5. Affordable and clean energy
6. Decent work and economic growth
7. Industry, innovation and infrastructure
8. Reduced social and economic inequalities
9. Sustainable cities and communities
10. Responsible consumption and production.

Annually, the UN World Tourism Organisation (UNWTO), World Travel and Tourism Council (WTTC), Pacific Asia Travel Association (PATA) and other transnational tourism associations award and recognise destinations and tourism enterprises which fulfil all or most of these UNSDGs. Many businesses have recognised that apart from the positive ethical considerations associated with sustainable tourism, a growing sector of the tourism market is attracted to sustainable destinations and to sustainable tourism experiences and products. In other words, sustainability sells and its increasing embrace by tourism enterprises reflects both the ethical and economic imperatives.

However, many tourism destinations and enterprises have fallen well short of achieving sustainability goals. Tourism in both developed and developing countries has frequently exploited destination communities and their resources of food, water, land and people. At its worst, tourism, with the corrupt collaboration of local authorities, has played a significant role in exacerbating economic and social inequalities and crime and has exploited local people as cheap labour for tourism enterprises. Destinations that ignore sustainability principles can sometimes find that despite what appear to be favourable key performance indicators (KPIs) (number of visitors, spend per capita, length of stay) tourism operates at a net economic loss for a destination. This is because most of the spending benefits offshore companies and businesses and little of the money spent on tourism in a given destination stays in the destination community. The term for this is leakage. High and sustained levels of leakage can potentially be a major risk for tourism enterprises and destinations. High levels of leakage can trigger a crisis for an affected destination (Weaver and Lawton, 2014).

Although the economic dimension of tourism can and should be viewed in its sustainability context, tourism from an economic perspective involves a range of transactions at both the business-to-business level and the business-to-consumer level. In 2019, the global tourism industry totalled over US$8.9 trillion of international and domestic transactions, accounting for 10.3 per cent of global GDP, generated 330 million jobs, and accounted for US$1.7 trillion in visitor exports (28.3 per cent of global service exports and US$948 billion in capital investment or 4.3 per cent of total global capital investment (WTTC, 2020)).

By 2019, tourism was a significant economic sector in practically every country in the world. During the Covid-19 pandemic of 2020, the WTTC estimated in November 2020 that between 140 million to 174 million people lost their jobs and that the tourism industry was likely to contract by between 43 and 53 per cent compared to 2019. International visitation was expected to drop between 65 and 73 per cent compared to 2019 levels (WTTC, 2020). The reason for the variations in forecasting are dependent on the longevity of the pandemic crisis. The best-case forecasts assume an instant end to the crisis, the worst case, its continuation. The WTTC tends to be the most optimistic of the global tourism bodies, which have extensive and research resources. As discussed in Chapter 4, there is a close link between economic and political stability.

Economic stability includes a number of key elements, which facilitate trade, financial transactions and tourism, and ensure both consumers and service providers benefit from the transaction and its value proposition.

This chapter includes a case study on the Global Financial Crisis (GFC) (2008–2009) and its impacts on tourism, and examines how the tourism industry and tourism destinations responded to and recovered from the GFC. A second brief case study examines the implications of the recent Greek economic crisis (2010–2017) on tourism to Greece.

KEY ISSUES IN THE RELATIONSHIP BETWEEN ECONOMICS AND TOURISM

A reliable and functioning banking system, steady rates of international currency exchange, and clear and consistent terms of trade are just three of the many KPIs of a stable economy. At a broader dimension, low levels of national unemployment and inflation are indicators of a well-managed economy. Economic stability is enhanced when interest rates and inflation are low and the majority of working people are paid liveable wages and salaries. Many advanced economies benefit from a government funded social security (services) safety net, which provides financial support for people who are unemployed, require medical care or have reached retirement age without resources of their own.

Between 2010 and 2019, global tourism grew rapidly due to significant improvements to the economic environment in many countries previously defined as developing economies. China, India, Japan, South Korea and many of the Association of Southeast Asian Nations (ASEAN) countries experienced a significant growth in the size of their middle classes, which had earned a growing portion of the population sufficient discretionary income to spend on travel and tourism. The growth of the Asian middle class coincided with the rise of low-cost carriers (especially in the Asia Pacific region), which led to reductions in the costs of many international air fares and enhanced the affordability of travel.

As Chapter 3 discusses in depth, most of the economic gains that tourism made between 1970 and 2020 were lost due to the catastrophic economic impact of the Covid-19 pandemic during 2020. The pandemic and its impact on tourism require a new strategic approach to rebuild global tourism's economic fortunes and compensate for the loss of trillions of dollars in business transactions and between 140 to 190 million jobs during 2020.

The globalised and heavily interlinked nature of most of the world's economy results in the global community sharing both the benefits and costs of global economic booms and busts. Many global economic booms and busts appear to be unrelated to economic activity in individual countries.

Case Study 9.1

The Global Financial Crisis 2008–2009

The Global Financial Crisis (GFC) of 2008–2009 was essentially a result of appalling, and in some cases, criminal mismanagement of the banking and housing loan system in the USA (Dwyer and Sheldon, 2010). The impact of sub-prime housing loans sold mainly to low-income earners or welfare recipients in the USA by a range of financial institutions (loans offered to people who had no capacity to repay them) spread through the global banking system and led to a run on global stock exchanges in 2008 and 2009. Sub-prime housing loans were usually offered at high interest rates to borrowers with little or no collateral. The high interest rates that applied to such loans acknowledged the high delinquency risk (Papatheodorou et al., 2010).

These dubious loans were in turn guaranteed by other banks and financial institutions, which rapidly added to their debts. In the USA, the Federal Home Loan Mortgage Company (Freddie Mac) and the Federal National Mortgage Association (Fannie Mae) were government-backed instrumentalities, charged with guaranteeing mortgage funds. Both they and many other financial institutions worldwide failed to observe due diligence on millions of dishonoured sub-prime US housing loans.

The absence of due diligence on the part of the banks, governments and multinational financial institutions in approving and guaranteeing these loans spread like a virus of dishonoured debt throughout the global banking system. The crisis in the USA was exacerbated in 2008 when the value of housing (the asset base of these sub-prime loans) fell precipitously, and the cost of foreclosure to lenders exceeded the market value of the property. In the USA, the major banking collapse was Lehman's Bank in 2008. During 2008, the final year of George W. Bush's administration, the US government was literally impelled to extend government guarantees to Freddie Mac and Fannie May, two of the most powerful financial institutions in the USA. This controversial decision, objected to by many conservative economists, arguably prevented a complete collapse of the US banking system.

The collapse of several banks and financial institutions in the USA had a negative impact on share markets and businesses all over the word. It meant that the ability of financial institutions to offer loans or extend credit to businesses was severely compromised. The GFC led to a loss of confidence in banking and financial institutions and a decline in global business and trade. Currency exchange rates experienced instability and economic uncertainty led to job losses in the USA and countries distant from it.

From an international tourism perspective, economic uncertainty and a global economic downturn is a major threat to tourism. Leisure tourism, in particular, involves discretionary expense. When people are confident about their job security and the stability of their income level, they are happy to spend money on travel. When they are uncertain about the reliability of their income and job security, they will either reduce travel expenditure or simply abstain from non-essential travel. These behaviours were clearly evident during the GFC between 2008 and 2009 (UNWTO and International Labour Organisation, 2013). Economics is a factor in every tourism crisis, irrespective of causation. Economic indicators feature as a measurement of tourism downturns and recoveries. Tourist numbers, spending patterns, revenue for tourism businesses and length of stay are the statistical measuring sticks of crisis and recovery. However, the state of the economy at a local, national, transnational or global level can precipitate a downturn or a crisis for tourism.

Globally, during the GFC, between 2008 and 2010 as indicated by Figure 9.1, there was a four per cent global decline in the number of international travellers but a far greater decline in tourism spending, as illustrated in Figure 9.2. While the declines were a problem for tourism, the downturn was less than a crisis but a clear disruption of a strong growth pattern in global tourism, which was evident between 2004 and 2007. A notable factor in tourism decline was that it emanated largely from developed countries and high-income travellers, which were the countries and socio-economic sectors most impacted by the GFC. Many countries (including Dubai and Abu Dhabi), which had focused on developing up-market tourism infrastructure, experienced a significant decline in demand for five-star accommodation. This resulted in many vast five-star hotels in Dubai and especially Abu Dhabi (which almost exclusively offered five-star accommodation), hosting few paying guests. This was reflected in a decline of visitor numbers for those countries that lacked a diversity of accommodation standards and price points.

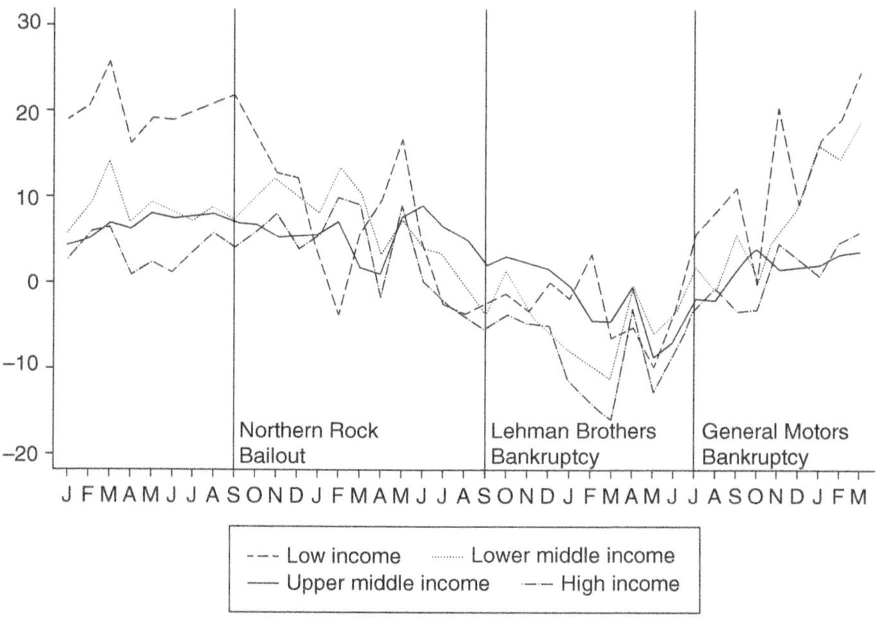

Figure 9.1 Declines in tourism numbers globally by national income levels during the Global Financial Crisis 2008–2009

Source: UNWTO and International Labour Organisation, 2013. Used with kind permission of the UN World Tourism Organisation.

(Continued)

In June 2009, 75 tourism academics gathered in Mallorca, Spain, for the International Academy for the study of tourism. They were there to discuss the implications on tourism and responses to the GFC. Two of the keynote presenters included the doyen/nes of global tourism economic research, Professors Larry Dwyer and Pauline Sheldon (2010). The gathering committed the scholars to undertake related research. While many did so, most subsequent academic research tended to apply to a single country. This research was useful but hardly global. An excellent paper by Papatheodorou et al., (2010) 'The Global Economic Crisis: Consequences and Perspectives', approached the issue globally.

For students and practitioners of tourism who seek to study the impact of the GFC on tourism in-depth, the ultimate source is a joint publication by the UNWTO and International Labour Organisation (2013) *Economic Crisis, International Tourism Decline and Its Impact on the Poor*. This book provides an extensive global analysis of the impact of the GFC on tourism and the industry's response. It also includes a number of case studies focused on developing countries.

During the GFC, there was a decline in long haul, long stay international travel but a growth in short stay, short-medium haul travel. Demand for first and business class airline seats declined and many carriers introduced a premium economy class section, which became popular among business travellers who sought extra comfort at a small price premium. Many businesses, which traditionally spent large sums of money on business-related travel, reduced their travel budgets and adapted their travel policies in response to a more uncertain and less profitable business environment.

The GFC led to a significant growth in the number of low-cost carriers and the patronage of low-cost carriers (LCCs). Many established airlines established low-cost carrier brands. The most prominent carrier to do so during the GFC was Singapore Airlines, which launched its LCC brand, Scoot in 2012. The unbundled LCC concept, which was highly successful for airlines, was extended to some hotel groups, including the Malaysian-based Tune Hotel Group. The concept was also extended to cruising when the Easy Jet company launched Easy Cruises in the Aegean in 2008 as a no-frills cruise operation. This experiment was a limited success. While the GFC represented a short-term crisis for global tourism, the tourism industry adapted to the decline in spending by offering a wide range of inexpensive product options.

As Figures 9.1 and 9.2 indicate, the GFC was far from an unmitigated disaster for tourism in developing countries. Travellers from the developed world were spending less money on their travel and those countries deemed as good value for money or to put it more bluntly, cheap, attracted slight growth in tourism arrivals and spending during the GFC while spending and arrivals in developed countries declined.

The overwhelming response of the tourism industry and destinations in response to the GFC would involve emphasising of the value proposition travel. As some Gulf states learned, to their immense cost, diversification of the tourism infrastructure and consumer offering should include a range of budgets, not just the high rollers. During

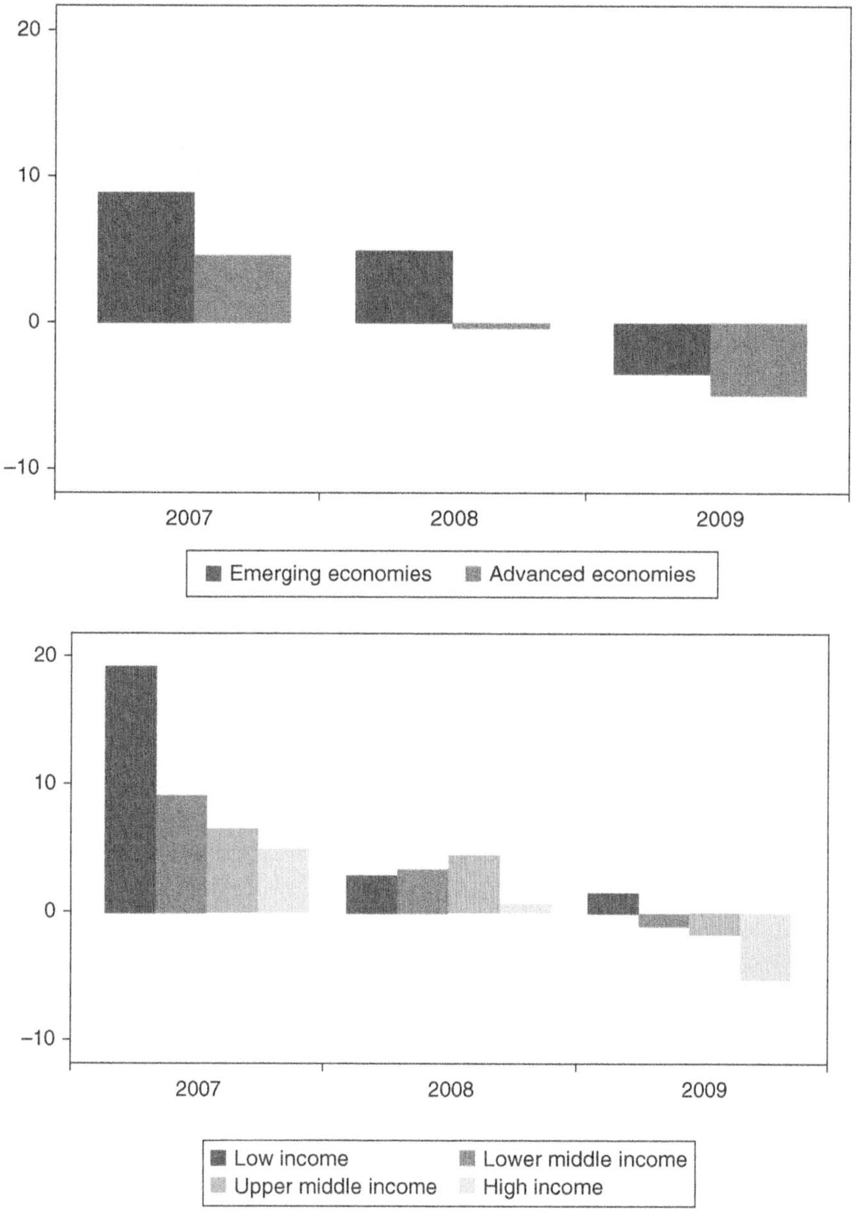

Figure 9.2 Changes in global tourism expenditure 2007–2009 for developed and developing countries, and wealth of tourists

Source: UNWTO and International Labour Organisation, 2013. Used with kind permission of the UN World Tourism Organisation.

(Continued)

the GFC, many high rollers lost their fortunes and companies, and some finished up doing time in penitentiaries for their contribution to the GFC.

The GFC of 2008–2009 and the Asian Financial Crisis of 1997–1998 (Henderson, 2007) are two examples of crisis events with an overwhelmingly economic dimension. As mentioned earlier, many crises discussed in this book include major economic implications. Natural disasters, pandemics, terrorism, political instability and high crime rates all have an economic impact on tourism and many other sectors of business. The economic impact of the 2020 Covid-19 pandemic on the global tourism industry has been catastrophic and most sectors of the global economy have experienced an economic recession as a consequence. Yet, in defining Covid-19 as a crisis, it is fundamentally a pandemic but a pandemic with significant economic implications. As discussed in Chapter 3, unlike previous crises which have altered the nature of travel demand and tourism flows, Covid-19 radically disrupted tourism as a business and leisure activity.

CURRENCY EXCHANGE RATES AND INTERNATIONAL TOURISM

A core economic issue that can and does affect tourism flows is currency exchange rates, which determine the relative costs of tourism services between countries. In the context of international tourism costs for consumers, tourism principals and operators are generally determined in hard currencies: US dollar, Australian dollar, pound sterling, euro, yen, or in the case of China, renminbi. Consumers in tourism generating regions will pay for their international tourism bookings in their local currency but exchange rates can be a significant determinant of demand for travel consumers in the tourism generating region and in the pricing of tourism services in destinations.

When most international tour operators (wholesalers) determine brochure prices for tourist product in their destinations they normally allow for an acceptable margin of movement in exchange rates between hard currencies and the currency in the travel and tourism generating region (TTGR). As more people buy tourism products online and more tour operators have e-brochures, it's much easier to adjust list prices than it was in the days of printed brochures. In cases where exchange rates fluctuate many operators indicate a price on application (Beirman, 2016).

When the exchange rate between a tourism generating region and the destination is favourable for the consumer, there is generally an increase in demand. When the exchange rate is unfavourable, this often results in a decrease in demand. In cases where the fluctuations in exchange rates clearly disadvantage a key source market, occasionally tourism businesses may actually reduce hard currency prices for that particular market if it is deemed valuable. This is certainly the case with air fares. Although in theory, international air fares are priced in accordance with a standard currency called a NUC (neutral currency unit based on the US$), in practice, airlines have a mission to fill planes and there are often significant differences in prices between city pairs depending on the departure point (IATA, 2018). At times, exchange rates and the state of the economy have deemed that buying a return air fare between London and Sydney has been far cheaper if purchased in London than in Sydney, and at other times it has been the opposite.

The euro was initially introduced as a common currency for European Union (EU) member states in 1999 although not all EU member states adopted it. The UK was the most prominent EU member state to reject adoption of the euro. It was notable that when the euro became widely accepted as a currency in many EU countries in 2000, prices for tourism-related goods and services escalated in Greece, which adopted the euro. This presented an opportunity for Turkey (a non-EU member with a weak currency) and Croatia, which was an EU member but had rejected the euro and had its own currency (kuna: 7 kunas = 1€ at time of writing). This led to comparable tourism products and services in Croatia being less expensive than in Greece. From 2000 to 2010, international tourism arrivals to Greece grew slowly

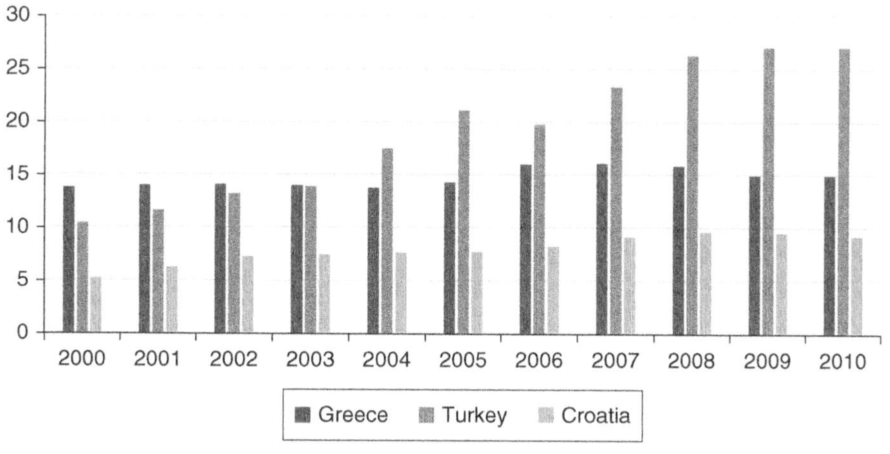

Figure 9.3 Comparison of international tourism arrivals to Greece, Turkey and Croatia 2000–2010

Sources: Baldigara and Mamula, 2012; Ministry of Tourism and Culture, Turkey, 2020; World Bank, 2020. Graph by the author

while tourism to Turkey and Croatia grew rapidly by comparison. Many of the qualities that attracted holiday visitors to Greece (sun, sea, sand, islands), were also present in Turkey and Croatia at relatively lower prices. The favourable exchange rates compared to the euro, which applied to Turkey and Croatia, contributed to stronger growth in tourism arrivals to Croatia and Turkey than to Greece during that period, as indicated by Figure 9.3.

As the case study on Greece in this chapter will show, Greece eventually overcame its exchange rate disadvantage after 2010 partially because of terrorism in Turkey 2015–2017 and partially because Greece's economic crisis of 2015–2016 forced a re-evaluation of tourism pricing in Greece to make the destination more price competitive.

Most international tourists to a greater or lesser extent are price conscious. This was especially evident during the 2008–2009 GFC. The cost of both transport to and from a destination and the cost of accommodation, food and tourism services within a destination are significant factors in destination choice, but far from the only factor. The quality of destination marketing plays an important role in stimulating demand. Referring to the comparison between Greece, Croatia and Turkey, the destination marketing of Greece was far less effective in key source markets than either Turkey or Croatia. The Greek government had an unfortunate habit of appointing managers of their overseas tourism offices who were political appointees rather than appointees who had anything vaguely resembling professional tourism marketing skills. By contrast, Turkey and Croatia were highly professional in their destination marketing (Pike, 2008).

One of the key reasons for Thailand's appeal as a destination until 2020 was that travellers from most of its source markets believed they got outstanding value for relatively little money. By contrast, the Scandinavian destinations and Iceland have a high level of experiential appeal but are very expensive and appeal to more affluent travellers.

In cases where there are significant exchange rate fluctuations, tourists may focus on those destinations deemed more affordable when a choice of common destination qualities is available. In some countries, consumer laws require transparent pricing especially in packaged or bundled tourism products and at times even demand the exchange rates on which the pricing was based. There have been some animated exchanges between the Council of Australian Tour Operators and the Australian Competition and Consumer Commission (ACCC) on these and other pricing issues. The ACCC expected that tips (which are optional) should be included in cruise pricing on brochures and advertising.

There is no doubt that in extreme cases of exchange rate fluctuations, the potential for radical changes in travel demand to a destination can occur.

ECONOMIC DEPENDENCE ON TOURISM IN DEVELOPING ECONOMIES – A MAJOR RISK

One of the basic maxims of economic management for business and governments is diversification – avoid over-dependence on a single market, a single product, service or a single customer. When times are good, this dependence can be a reliable source of income but the clear risk is that over-reliance on a dominant source market, product or service is a source of economic vulnerability. Oil-dependent economies in the Middle East and Russia have experienced economic pressures in recent years when the global demand price of oil and gas fell, as was the case in recent years. The same has applied to countries where the economy is based on the production and export of a dominant commodity or a few commodities, subject to demand and price fluctuations. Dependence on a single source market can be a major problem if that source market dries up due to economic problems within that market or due to political disputes that arise between the countries.

Political tensions between Western countries and China, exacerbated by Covid-19 travel restrictions, have led to China, which in 2019 was the major tourism source market for many Asia Pacific countries, to suspend international outbound tourism during most of 2020. Between 2010 and 2019, China became the major source of tourism for many destination regions. These included: Hong Kong and Macau (both Special Autonomous Zones of China), South Korea, Japan, Thailand, Malaysia, Singapore, Indonesia and Australia (to name a few). The tourism reliant economies of many Caribbean countries that have traditionally been reliant on tourism from USA, Canada, UK and Western Europe have experienced severe downturns in inbound tourism during 2020 due to Covid-19, and this has led to business closures and unemployment on a significant scale. Figure 9.4 indicates the loss of tourism jobs as of April 2020. By the end of 2020 the numbers indicated in the figure have almost doubled. According to the World Travel and Tourism Council, 62 million tourism jobs were lost during 2020.

Within Europe, Spain and Greece, two EU countries in which high levels of national debt were a major economic concern and which were heavily reliant on tourism for export income, have experienced severe economic strain during 2020. Relevant to this chapter, these two countries led campaigning for EU-based travel corridors as a means to maintaining tourism flows during the Covid-19 outbreak.

As figure 9.4 indicates, the Asia Pacific region accounted for some 60 percent of COVID-19 related tourism industry job losses during 2020. Many of the countries with significant job losses were developing countries in South-East Asia, including Thailand, Cambodia, Myanmar, Vietnam and South-West Pacific island nations,

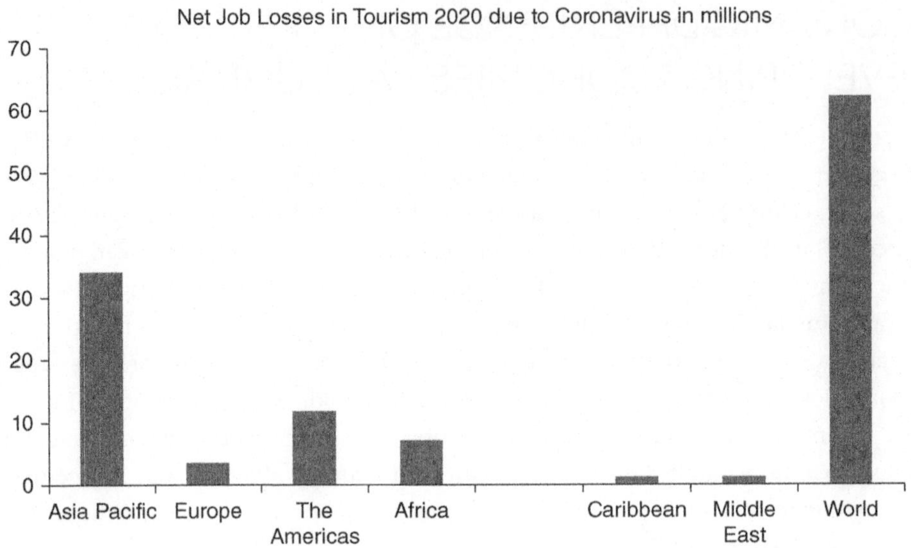

Figure 9.4 Global Job losses in tourism during 2020 due to COVID-19 by Region

Source: World Travel and Tourism Council 2021. Diagram by author with kind permission of the World Travel and Tourism Council. https://wttc.org/Research/Economic-Impact

including Fiji, Vanuatu, Tonga, Tahiti and the Cook Islands. Most of the world's highly tourism dependent economies are developing countries (Avraham and Ketter, 2016). This applies to the Asia Pacific region, Africa, the Caribbean and central and Latin American countries, including Mexico, Colombia and Cost Rica. In the developed world, many tourism professionals who have either lost their jobs or their employment have been suspended or supported by government social welfare schemes. These economic safety nets do not exist in the majority of developing countries. In developing with tourism dependent economies, the loss of tourism exacerbates poverty and the social problems that accompany poverty, including crime.

In addition to job losses, tourism-related infrastructure has rapidly deteriorated due to the lack of income to fund maintenance of hotels, airports and roads or capital infrastructure, including aircraft, coaches and resorts that are left unused and largely unsaleable. Assets literally become liabilities. The loss of tourism has its own reverse multiplier effect in that many non-tourism enterprises ranging from agricultural production, construction, trades, local transport services, retail outlets and markets, to cafes, restaurants and bars lack the income to remain viable.

Many developing countries have experienced crisis events, which have negatively impacted on tourism prior to 2020, but the extent of economic damage inflicted by Covid-19 has driven home the lesson that countries that have traditionally relied on tourism need to diversify their economies.

Case Study 9.2

The Greek Financial Crisis and Tourism to Greece

Greece is one of the world's most varied and attractive tourism destinations. Blessed with beautiful scenery, a long and sunny summer, over 3,000 islands dotted within the Aegean Sea and a long and fascinating ancient and modern history, Greece has much to attract tourists. The romance of tourism to Greece has been popularised by movies, including *Boy on a Dolphin, Zorba the Greek, Shirley Valentine* and *Mamma Mia* 1 and 2. As the 'cradle of Western civilization', Greece's many architectural masterpieces, outstanding works of art, cultural treasures, remnants of Minoan, Mycenaean and the glory of Athenian civilisation during the age of Pericles (500 BCE) attracted tourists well before the era of mass tourism. Greece has an excellent tourism infrastructure on the mainland and the many inhabited islands. The Greek tourism industry hampers many opportunities for growth due to the long off-season closure of many of its tourism operations between November and April. Greek tourism policy makers have paid little attention to flattening out seasonality fluctuations by an effective cool weather tourism marketing campaign. The Greek agricultural sector grows and produces a wide range of excellent food and wines, and its people are hospitable. In many respects, Greece is literally built for tourism.

Of all the EU countries, Greece has the highest economic reliance on tourism. By 2019, 20.8 per cent of the Greek national economy and 12.1 per cent of GDP was based on tourism. Tourism is a major employer for the country, employing 846,000 people or nearly 20 per cent of employed Greeks (WTTC, 2020). As discussed earlier in this chapter, the relatively reluctant entry of the Greek government and economy into the Eurozone currency group placed Greece at a competitive disadvantage with two of its key Eastern Mediterranean regional competitors for the sun, sand and sea leisure market – Turkey and Croatia between 2000 and 2010. Although Greece successfully hosted the 2004 summer Olympic Games, enhanced transport infrastructure in Athens and basked in the media attention, it never succeeded in

(Continued)

attracting the level of follow-up tourism the Greek government hoped to generate from hosting the Games. This problem is not unique to Greece. Many recent Olympic host cities, including Atlanta 1996, Sydney 2000, Athens 2004, Beijing 2008, London 2012 and Rio de Janeiro 2016 have been disappointed by the lack of a post-Olympic tourism spike. The Greek tourism industry did itself no favours by attempting to grossly inflate tour and hotel prices prior to the 2004 Olympics, only to find the demand on the part of tourists to be 'fleeced' was minimal and many of these overpriced hotels were empty until market forces led to significant price reductions to fill the hotels (Papatheodorou and Arvanitis, 2014).

Nevertheless, as Figure 9.5 demonstrates, between 2012 and 2019 inbound tourism to Greece enjoyed an impressive rate of growth, doubling in that period. This growth occurred despite the Greek government being forced to deal with an austerity programme enforced on it by the EU in response to a massive accumulation of government debt. The surge of tourism to Greece was partially a result of enhancements to tourism marketing and infrastructure but also a response to the travails of many of Greece's main competitor destinations during that period. Uprisings and civil war in Tunisia, Libya, Egypt and Syria from the late 2010, some of which continued throughout 2020, terrorism targeting tourists in Turkey 2015–2017 and growing anti-Christian sentiment in Turkey have been major contributing factors to the declines of tourism in those countries and the surge in tourism to Greece until the 2020 outbreak of Covid-19.

The adoption of the euro currency by Greece in 2000 was a dampener to the growth of Greek inbound tourism between 2000 and 2007. Turkey, Croatia, Bulgaria, Egypt, Tunisia, Syria (you read right) and Jordan gained significant growth at the expense of Greece, which was regarded by many tourists who sought an Eastern Mediterranean sun, sand, sea and history/culture destination as grossly overpriced, largely due to its product being priced in euros.

During the GFC (2008–2009), Greece experienced a slight downturn in tourism arrivals but a steeper decline in tourism spending and length of stay in line with many European destinations in which the luxury travel market contracted significantly. Severe pressure was imposed on the Greek economy to act on austerity, applied by the EU on the Greek government between 2008 and 2018. The level of Greek government debt rose to unsustainable levels and the EU demanded a period of economic austerity on the part of the Greek government and a timetable for debt repayment. These demands were not without reason. In Greece, tax avoidance was a national sport, social service payments were generous, productivity was among the lowest in Europe, the civil services were a refuge for the inept, governance was conducted by the incompetent and retirement ages were low by EU standards. The aforementioned comments are not the opinion of the author but the thoughts of hundreds of thousands of hardworking and largely successful Greek immigrants to Australia who left Greece to escape the country's social, political and problems (Price, 1975).

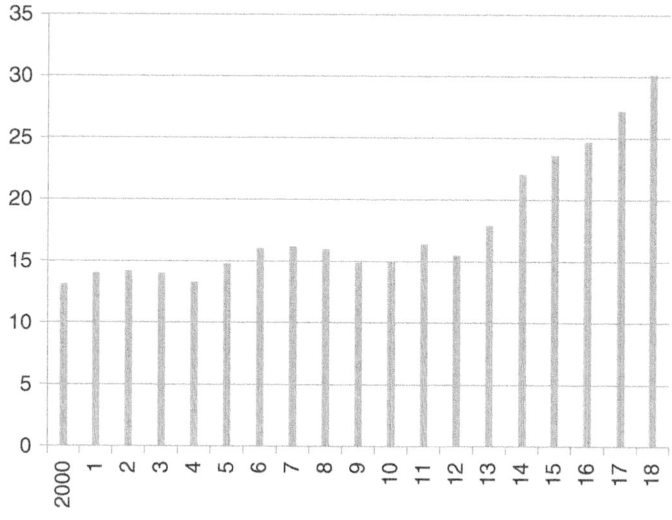

Figure 9.5 International tourist arrivals in Greece 2000–2018 (millions)

Source: Greek Ministry of Tourism, 2019. Graph by the author

The calls for austerity by the EU were resisted by the Greek government and sections of the population who accused the EU's position as being dictated by Germany, and as a result, this was seen by some Greeks as a neo-Nazi attempt to control Greece. In objective economic terms, the EU had little alternative but to step in and force Greece to adopt responsible economic practices to save Greece and ultimately the EU from bankruptcy. To put it in perspective, Greece was not alone in the EU as a major debtor. Spain and Italy were also on the EU's high debtors hitlist but the Greek economy was the most exposed.

Between 2011 and 2017, there were extensive and occasionally violent outbreaks of civil unrest and protests in Greece directed at both the EU and successive Greek governments, which had implemented many of the severe EU austerity demands. Despite the dramatic media coverage by the tabloid media, which tried to depict Greece as a country in flames, nearly all the anti-government demonstrations took place in Athens, most within 500 metres of Syntagma (Constitution Square), where the Greek parliament and the Presidential Palace are located (Rodosthenous, 2017). A major challenge for the Greek Ministry of Tourism and hundreds of their tourism industry stakeholders and millions of travellers around the world was to communicate the real extent of unrest and counter the lurid reporting of the tabloid media on all platforms. Between 2009 and 2013, the Greek Ministry of Tourism hosted tour operators servicing Greece from all key source markets. They also hosted travel agents and travel writers from all media platforms to experience Greece first hand

(Continued)

and see that apart from the political district of Athens, tourism was running normally and the island destinations were safe. The Greek Ministry of Tourism significantly enhanced its social media presence.

The significant growth of international tourism to Greece between 2011 and 2018 was partially due to the enhanced marketing on the part of the Greek Ministry of Tourism and the tourism industry. More influential was that many of Greece's destination competitors succumbed to uprisings, civil war and terrorism. The misfortunes afflicting Turkey, Syria, Lebanon, Egypt, Libya and Tunisia led to tourists seeking their Eastern Mediterranean sun, sand, sea and culture experience in Greece, Cyprus or Israel. Despite the resurgence in tourism to Turkey from 2017, Greece emerged as a favourite Eastern Mediterranean destination until the 2020 outbreak of Covid-19. Ironically, despite Greece experiencing a severe economic crisis, between 2008 and 2017, tourism to Greece thrived, largely due to the fact that their economic problems were greatly exceeded by the far more severe political and social problems of many of its neighbouring countries. The growth of Greek tourism also aided the Greek economy in making some headway towards addressing its economic crisis.

SUMMARY AND CONCLUSION

Economic issues play a significant role in influencing demand for tourism globally. As leisure tourism especially involves discretionary expenditure, the state of global and national economies determine the number of people who can afford to travel domestically and internationally. Economic downturns don't necessarily translate into major reductions in tourism activity but the nature of that activity may change in relation to destination choice, length of stay, per capita spend and the travel choices people make in relation to transport, accommodation type, tour options, etc.

As the GFC demonstrated, certain sectors of tourism, notably business travel, can undergo major changes in spending patterns during times of economic hardship. The GFC forced many business travellers to abandon first and business class travel in favour of premium economy or economy travel and downgrade accommodation from five-star hotels to three- or four-star hotels. Business travel managers adopted tighter controls on travel-related expenditures. The GFC coincided with a boom in the patronage for and a surge in the number of low-cost carriers to meet the growing business and leisure demand for budget travel.

For destinations, flexibility in tour product and service choices and variety of tourism infrastructure catering to multiple budgets is a vital risk management strategy during periods of economic instability. Pitching a destination exclusively for high spending luxury travellers proved to be a dangerous strategy during an economic recession. The United Arab Emirates discovered this to their cost during the GFC but were quick to diversify their accommodation infrastructure after the GFC. Another significant economic imperative involves ensuring that destinations and tourism business operate in a sustainable manner that ensures tourism spending in a destination as subject to minimal leakage economically, and benefits the destination's economy and community. Observance of UN sustainability principles is designed to ensure that tourism avoids the economic and social trap of being an industry that exploits the destination community as a source of cheap labour.

The tourism industry needs to be aware of day-to-day economic issues when engaging in destination or business planning. These include accounting for exchange rate fluctuations, competitive price matching and yield management. In recent years, the growth of demand for low-cost carriers, accommodation disrupters such as Air BnB and transport disrupters such as Uber have actually contributed to a growth in the tourism industry between 2010 and 2020. As discussed in Chapter 3, Covid-19 has forced all tourism businesses to re-evaluate the traditional high volume, low yield basis of doing business.

Until the outbreak of Covid-19 during 2020, many countries, especially developing countries, regarded tourism as a natural pathway to economic advancement and employment creation. Important as tourism has been to the global economy, no country should be over-dependent on any single economic sector. While tourism will continue to be a significant contributor to economies in the future, the 2020 pandemic has demonstrated that global tourism is vulnerable to a major crisis event and that resilient economies require multiple sources of income, economic activity and expertise.

Discussion Questions and Issues

1. Pick a country of your choice and discuss the impacts that the GFC had on tourism to your chosen country.
2. The growth in demand for low-cost carriers is clearly a consequence of the GFC. Why do you feel it maintained growth after the GFC and in the post-Covid-19 period? Do you think demand will remain strong? Discuss in groups.
3. Greek tourism continued to grow despite the country's broader economic crisis. Do you agree or disagree with the author's assertion that growth in tourism increased to Greece largely due to crises affecting some of its competitor destinations, and if so, why?
4. Choose a destination country with a high level of economic dependence on tourism. Outline what steps you, as leader of the government, would undertake to diversify the economy given the country's natural resources and skills.

REFERENCES

Avraham, E. and Ketter, E. (2016) *Tourism Marketing for Developing Countries*. Palgrave Macmillan: Basingstoke, UK. pp. 9–23.

Baldigara, T. and Mamula, M. (2012) 'Tourism statistics in Croatia. Present status and future challenges', *Proceedings of Social and Behavioural Sciences*, 44: 53–61. XI International Conference on Service Sectors in Terms of Changing Environment.

Beirman, D. (2016) (Updated 2019) *Tourism Risk, Crisis and Recovery Management Guide*. Council of Australian Tour Operators: Sydney, Australia.

Dwyer, L. and Sheldon, P. (2010) 'The global financial crisis and tourism: Perspectives of the Academy', *Journal of Travel Research*, 49(1): 3–4.

Greek Ministry of Tourism. (2019) http://visitgreece.gr Accessed 15 May 2021.

Henderson, J. (2007) *Tourism Crises, Causes, Consequences & Management*. Elsevier: Burlington, MA, USA. pp. 17–29.

Interational Air Transport Association (IATA) (2018) Passenger Glossary of Terms. https://www.iata.org/iata-passenger-glossary-of -terms.xlsx. Accessed 20 December 2020.

Ministry of Tourism and Culture Turkey (2020) *Tourism Statistics*. https://ktb.gov.tr/EN-249283/tourism-statistics.html. Accessed 9 December 2020.

Papatheodorou, A. and Arvanitis, P. (2014) 'Tourism and the economic crisis in Greece: Regional perspectives', *Région et Développement*, Vol. 39: 183–203.

Papatheodorou, A., Rosello, J. and Xiao, H. (2010) 'Global economic crisis and tourism: consequences and perspectives', *Journal of Travel Research*, 49(1): 39–45.

Pike, S. (2008) *Destination Marketing. An Integrated Marketing Approach*. Elsevier: Amsterdam. pp. 70–99.

Price, C.A. (1975) *Greeks in Australia*. ANU Press: Canberra, Australia.

Radosthenous, N. (2017) *The Impact of the Economic Crisis on Tourism. The Case of Greece*. Honours Thesis, Laurea Institute of Applied Science: Helsinki, Finland.

United Nations (2020) *Sustainable Development Goals*. https://un.org/sustainabledevelopment/sustainable-development-goals/. Accessed 10 December 2020.

UN World Tourism Organisation and International Labour Organisation (2013) *Economic Crisis, International Tourism Decline and its Impact on the Poor*. UNWTO: Madrid.

Weaver, D. and Lawton, L. (2014) *Tourism Management* (fifth edition). John Wiley & Sons: Milton, Queensland, Australia. pp. 226–45.

World Bank (2020) *Tourism Arrivals: Greece 2001–2018*. https://data.worldbank.org/indicator/ST.INT.ARRVL?locations=GR. Accessed 8 December 2020.

World Travel and Tourism Council (WTTC) (2020) https://wttc.org. Accessed 10 December 2020.

10

THEME 7: TECHNOLOGICAL RISK, CRISIS AND RECOVERY

INTRODUCTION

Advances in technology have played a significant role in the growth of tourism over the past 200 years. Technological changes in transport, tourism attractions, event and entertainment infrastructure, booking, ticketing and customer data storage have revolutionised tourism and reduced costs for business and consumers alike. This chapter includes one case study focusing on cybercrime, which is one of the major technological threats to all tourism businesses.

Technology, despite its many benefits and risk management backups, does fail occasionally, sometimes with catastrophic consequences. Aeroplanes crash, ships sink, trains derail, amusement park rides occasionally fail, and buildings catch fire or collapse due to structural defects. Computer systems and data are sometimes hacked, compromising financial transactions and customer data. Statistically, the likelihood of these risks occurring are relatively low but the consequences for those affected and for the reputation of the businesses involved are severe. Most travel and tourism businesses and industry sectors have developed risk management strategies and protocols to minimise technological failures, applying to specific tourism industry sectors and businesses. However, on those rare occasions in which technology fails or confidence in commercial data is compromised, the consequences for businesses involve a major challenge to prevent a reoccurrence and restore business reputation and customer trust.

In recent years, dependence on technology within the tourism industry has vastly increased. The process of booking travel was dominated by labour intensive, manual procedures until the mid-1980s and the early online booking systems were quite complex. Today, the process of travel bookings, confirmations and documentation is now largely achieved via global distribution systems (GDS), dominated by three multinational companies, Amadeus, Sabre and Travelport (Galileo, Worldspan and Apollo) (Amadeus, 2020), which link airlines, rail trips, hotels, tours, events and attractions with travel agents or direct corporate clients. In addition to facilitating bookings, these GDS systems facilitate a wide range of back-office functions, including booking records, business accounting and customer databases for booking agencies. However, even GDS companies and Air BnB, despite their strong cyber defences, have been subjected to data theft on numerous occasions (Hern, 2016; Whittaker, 2019).

A key impact of the technological advances in travel bookings has been that many staff members whose role was the processing of bookings in all sectors of the tourism industry have been replaced by IT technology. In most cases, they have been reassigned to jobs that deal directly with customers. Nevertheless, as a proactive risk management strategy many travel companies do train staff members to be in a position to manage bookings in the event that the technology breaks down. While most booking technologies incorporate backup systems, on those rare occasions when an airline, hotel or an event booking system crashes, a backup of 'old' skills can be invaluable.

Technological advances have led to significant changes in travel distribution. Traditional travel agents once handled the overwhelming majority of travel transactions. Today, the role of traditional high street travel agents is now being challenged by online booking platforms, dominated by the multi-branded Expedia group, which acts as a travel booking platform and includes a number of online accommodation booking brands. Travellers who book hotels via the Trivago brand view a choice of hotel prices all provided by Expedia online booking brands.

Important as technological risks are to tourism businesses, there has been relatively little serious academic research into the issue. Joan Henderson (2007) is the most prominent of the relatively small number of tourism scholars who has paid serious attention to the technological dimension of risk and crisis. Other academics including Sigala (2012), Pennington-Gray et al. (2013) and Park et al. (2019) have conducted excellent research in highlighting the contribution of social media and IT to crisis communication and crisis management in tourism. The travel industry itself takes the issue very seriously and high-tech sectors of the industry, including aviation, cruising, high speed rail and many large global hotel chains, convention centres interactive attractions managed by governments, the Disney Corporation and Merlin Entertainments, have developed complex protocols to protect and maintain technology in their sectors. The one case study in this chapter focuses on the outstanding research conducted by Skye Theodorou (2017) whose honours thesis for the University of Technology Sydney (UTS) enabled the New South Wales (NSW) Government's Department of Small Business to develop a strategic approach to small business risk management of cybercrime. As well over 90 per cent of tourism businesses are small to medium enterprises, this research was of great value and its embrace by government was a testament to its value. While Theodorou was undertaking her honours year, she was working for the NSW Department of Small Business.

TECHNOLOGY AND CHANGING PATTERNS OF TRAVEL DISTRIBUTION AND FINANCIAL TRANSACTIONS

Principals, including airlines, hotels and tour operators, are able to sell their products direct to customers via their online booking sites. Direct online selling eliminates the payment of commissions to intermediaries and creates options for principal service providers to reduce costs for consumers. Many principals, including hotels, airlines and attractions, offer a range of financial and non-financial incentives to consumers, encouraging them to book airline seats, accommodation and attractions admission directly. Many low-cost carriers insist that they only accept direct bookings, and if they do accept bookings made by intermediaries such as travel agents the intermediaries are not paid commission. Travel agents compensate for this by charging a service fee for clients who book non-commissionable travel services.

Tour operators (wholesalers) traditionally (until 2010) relied on travel agents as the sole source of their bookings. This worked well for travel agents and wholesalers alike. Selling a tour to meet the specific needs of a traveller, a couple, a family group or a group of friends can be a complex and extended negotiation. Wholesalers were more than happy to leave that to the travel agent who could contact the wholesaler online or by phone and request a specific tour programme. However, significant advances in online marketing and product descriptions have made it easier for travellers to make their choices and deal directly with the wholesaler.

While the majority of tour wholesalers obtain most of their business from travel agents to whom they pay commissions, there are a growing number of younger, tech savvy travellers who prefer to book direct. In 2019, Contiki Holidays, which specialises in the 18 to 35 age bracket, obtained 50 per cent of its bookings direct from clients (Young, 2019). Travel corporation tour brands such as Trafalgar Tours and Insight Vacations, which cater to the over-45-year-old market, still attract about 80 per cent of their bookings via travel agents. Older clients tend to value their relationship with travel agents more than younger clients. Cruising companies, which attract a significant cohort of 'senior' clients, receive 75 per cent of their bookings via travel agents.

In recent years, online travel booking platforms have emerged as a major competitor to traditional travel agents and wholesalers. As mentioned earlier, Expedia's multi-branded platforms perform many of the functions of travel agents such as the booking of flights, accommodation, tours, car rental, rail passes and events.

The volume of business transacted by Expedia brands exceeds most travel agency groups. In many instances, the consumer price of the travel product it sells to the consumer is highly competitive with travel agents. The major risk to both consumers and Expedia's reputation is that once a financial transaction has been finalised, follow-up contacts, adjustments to bookings or cancellations either incur a fee or are difficult to undertake. This is a service weakness that Expedia is in the process of addressing, especially for corporate travel clients who often need to make multiple adjustments to travel arrangements prior to and after departure.

Advances in information technology (IT) has enabled travel companies to build extensive data profiles on customers, enabling them to provide and offer customised products and services to match the consumption patterns and preferences of their new and repeat clients. On the rare occasions when IT systems crash, or are subjected to cybercrime or hacking, few companies can readily adapt to manual backup systems. While technology has reduced the number of people involved in processing travel and hospitality transactions, when technological systems do fail there are insufficient staff to manage the manual processing of bookings.

A major risk applicable to most tourism businesses involves online financial transactions. In fact, today, the overwhelming majority of travel-related purchases are transacted online through credit or debit cards of electronic funds transactions. Cash or cheque transactions are more common in developing countries than in the developed world or for small over-the-counter transactions. The travellers' cheque, which was pioneered by the London Credit Exchange in 1772 and made globally famous by Thomas Cook in the 1874, largely became redundant during the 21st century.

The risks for online transactions centre on cybercrime, dominated by the scamming or skimming of online transactions. Data theft includes vendor and client account details and the transfer of funds to the criminal. Although cybercrime is relevant to issues covered in Chapter 8 (crime), it can't be completely separated from technological risks. The ability to protect data, including financial transactions, is a vital business requirement in an increasingly online business environment. From a legal perspective, the travel company's duty of care includes the security of financial transactions.

TOURISM AND TRANSPORT TECHNOLOGY BEYOND THE COMPUTER, LAPTOP AND SMARTPHONE

The era of mass tourism after World War II led to transport technology growing bigger and more complex than ever. At the turn of 2000, a large cruise ship carried just over 2,000 passengers and 1,000 crew. This is similar in scale to the ill-fated Titanic, launched in 1911 only to sink on its maiden voyage, in 1911. By 2019, the world's largest cruise ships, Royal Caribbean Line's *Symphony of the Seas* could accommodate over 6,800 passengers and over 2,000 crew (CLIA, 2020). The cruise industry has strict protocols regarding maintenance of ocean-going ships, which are established by the International Maritime Organisation (IMO, 2020).

The introduction of the Boeing 747 jumbo jet in 1970 more than doubled the passenger capacity of commercial jet airlines from 189 (maximum capacity of the Boeing 707) to over 400. In 2006, the Airbus A380 increased passenger capacity even more. In mixed-class configuration the aircraft could comfortably hold 500 passengers and in economy only configuration, over 800 passengers. These changes enabled significant economies of scale and multiple price point options for cruise ship and airline operators. They have also added a new level of technological complexity and accompanying technological risks. Figure 10.1 provides an illustration of the growth in size of transport providers.

A civil aircraft is an incredibly complex piece of technology with millions of parts, complex computer controls and numerous safety backup systems. The world's two largest civil aviation manufacturers, Boeing in the USA and the British/EU consortium Airbus Industries, exhaustively test every aircraft they build before delivery to airline companies. The International Civil Aviation Organisation (ICAO) has volumes of strict maintenance and safety protocols applicable to every civil airline in the world. During 2019, the Boeing company was forced to ground all 737 MAX aircrafts as a result of two crashes, one in Indonesia and one in Ethiopia, due to a faulty part. The cost of

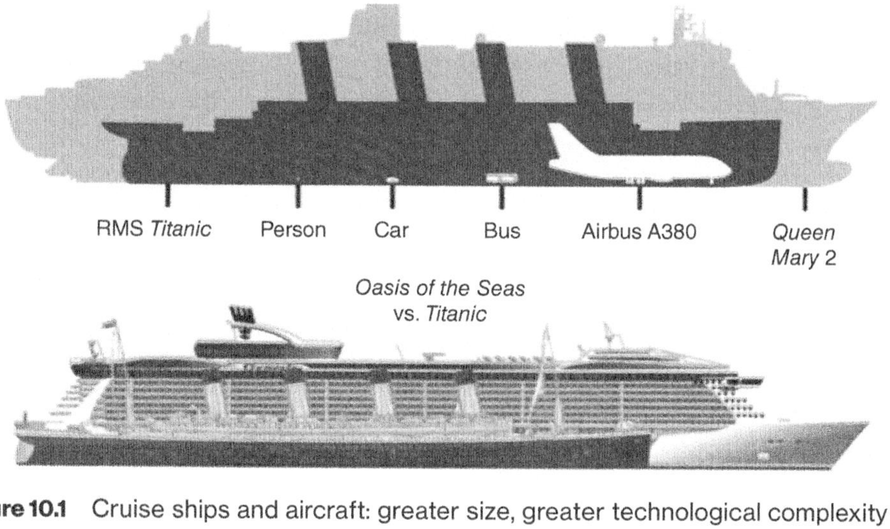

Figure 10.1 Cruise ships and aircraft: greater size, greater technological complexity

Source: Melvin Porter, 2018.

the two crashes was several hundred lives. Although air crashes in recent years have been very rare, the few which do occur exact a high toll in human lives and have a significant impact on the reputation of the airline involved. Airlines, for all their technological and safety measures, are vulnerable to natural phenomena, including wind shear, volcanic ash clouds and violent storms. In April 2010, airlines throughout most of Western Europe were grounded for a week when the risk of volcanic ash clogging jet engines forced the cancellation of thousands of flights to, from and within Western Europe (Beirman, 2012).

On land, railway travel experienced a significant revival in recent years due to the expanding global network of high-speed rail. Railway travel was exceedingly popular for tourists in Europe, Asia, the Americas and Africa during the 19th century and the first half of the 20th century. However, advances in aviation and the reduction of air fares between 1970 and 2019 imposed competitive pressure on railways and other forms of land transport. In 1964 (the year of the first Tokyo Olympic games) Japan launched the first high-speed rail service, the Shinkansen or bullet train, which travelled at over 200 km per hour between Tokyo and Osaka. Since then, high-speed rail services operate on many routes in Japan, the technology has been upgraded and there are widespread high-speed rail networks in Japan, South Korea, Taiwan, China, UK and Western Europe. Many of the services have average speeds well in excess of 300 km per hour and the CBD-to-CBD routes make them time competitive with short-haul flights, and competitive in term of comfort and price. Unlike airlines, there are few baggage limits for rail passengers. The maintenance of high-speed railways and the track that supports them is a massive undertaking. Countries with extensive high-speed rail networks, including France, Spain, Germany, the UK, China, Japan and the Republic of Korea (South Korea) operate massive maintenance depots where the rolling stock is thoroughly checked at regular intervals, usually monthly. Trains operating on maglev (magnetic levitation) routes, which enable speeds of over 400 km per hour, are checked even more frequently. Technological failure resulting from faulty systems in the train, the state of the tracks or en-route signalling could have catastrophic consequences.

Motor vehicles, especially those designated as tourist vehicles, also require strict risk management and maintenance guidelines whether they are a taxi, car rental vehicles or vehicles used to transport tour groups on or off-road. Vehicle maintenance provisions in most legal jurisdictions and driver skill requirements for tour operators are far stricter than for private vehicles and drivers. This results in expensive licence and registration fees. Uber, as a transport disruptor, employed or contracted private vehicles and their drivers to transport tourists, employing a legal loophole enabling minimal licences fees. This loophole is gradually being closed in many jurisdictions after legitimate tour vehicle operators and taxi drivers objected vociferously to being undercut by drivers who were not subject to tour vehicle or professional driver licence fees. Motor vehicles, especially tour coaches and customised off road vehicles, employ increasingly complex technologies for performance over variable terrain and enhanced comfort, safety, wi-fi, and in-vehicle entertainment for tour passengers.

Mass tourism and large-scale infrastructure has not been limited to transport. The First World Hotel, located in the Genting Highlands, Malaysia, with 7,351 guest rooms, is currently the largest hotel in the world. Several hotels in Las Vegas, including the MGM Grand Hotel, approach or slightly exceed 7,000 rooms. The vast size and guest capacity of these and many other hotels and resorts reflect the massive growth in tourism demand in recent years. Many cities around the world have developed large convention, event and exhibition centres catering to well over 10,000 attendees. These business convention centres are integral to attracting the growing global market for business and cultural events and are often owned, supported or sponsored by governments. Infrastructure on this scale incorporates a wide array of technologies, which come with their associated risks. One is the assurance of backup power in the event that the public electrical supply fails.

Integral to the effective functioning of accommodation, attractions and event infrastructure is a complex web of integrated technologies. These include:

- Air conditioning and climate control
- Lighting and power
- Computer and wi-fi linkage
- Audio visual systems
- Staging technology
- Plumbing
- Food preparation and food and beverage storage technology
- Security technology, including CCTV
- Reservations, financial transaction and customer guest data
- Check in and checkout procedures and high-tech hotel keys
- Fire prevention, detection and sprinkling systems
- Emergency evacuation alarms and evacuation paths
- Elevators and escalators
- Computer-based games and gaming equipment
- Luggage and equipment control and identification technology
- Strict risk management certification for external contractors involved in staging, setting up exhibitions, lighting or temporary structure associated with events.

This list is expandable but it also indicates the technological complexities involved in delivering a service to guests and attendees of accommodation providers and event venues. Each of the technologies associated with managing events includes their own set of risks. Unlike cruising and aviation, there are no global risk management standards applying to hotels. However, the highly competitive and global nature of the hotel industry has led to most major hotel chains introducing strict risk management protocols applying to all hotels in the chain. Even in the case of multi-branded hotel chains such as Accor, which has over 20 hotel brands ranging from three-star Ibis Budget to the five-star Sofitel and Fairmont brands, are all required to maintain a common standard of risk management practice, much of which applies to the hotel's technology.

Attractions, especially interactive attractions, are increasingly dependent on high maintenance and high-tech elements. When Walt Disney opened the first Disneyland in Anaheim, California in 1955 (Gennawey, 2013), many of the attractions involved mechanical animals, high tech rides, a monorail and staging and lighting effects, which were state of the art at that time. They all required a significant amount of maintenance and quality control. The Anaheim Disneyland would be the first of an incredibly successful global franchise that would lead to the establishment of Disney theme parks in locations as varied as Orlando, Florida, Paris, Hong Kong, Tokyo and Shanghai. Disney resorts operate under a very strict and globally enforced set of risk management and maintenance protocols usually exceeding government requirements in the countries in which they operate.

The British based Merlin Entertainment (2020) operate well over 200 attractions globally, including a growing range of over 30 Madam Tussauds wax museum outlets worldwide. In common with Disney attractions, many of Merlin Entertainment's attractions are interactive and high tech, and they have a comprehensive range of risk management protocols to protect visitors and staff.

Amusement and theme parks are designed to entertain and thrill attendees of all ages but there have been some tragic disasters, and some of these linked to technological failure. In Australia, the Dreamworld theme park in Queensland's Gold Coast earned national and global notoriety in October 2016 when four people died on one of the feature rides, the Thunder River Rapid Ride. The coroner's report on this incident, released in 2020, found that the ride was poorly maintained and that the operator of the ride on that day was an unsupervised novice on his first day of work. From once being one of Australia's most popular theme parks, Dreamworld has experienced a long struggle to restore its damaged reputation and that of its parent company, Ardent Leisure. Central to this and other similar incidents is the duty of care of attraction operators. Anthony Cordato (2017), one of Australia's leading tourism lawyers, has frequently stated that conditions of entry written by attraction owners, claiming 'all care and no responsibility' for any accidents, are overridden in law by the obligation of duty of care, and adherence to occupational health and safety laws, which involve proper maintenance, operation and supervision of amusement park rides.

A WORD ABOUT SPACE TOURISM

Whenever a tourism conference includes a presentation on space tourism, it arouses a great deal of interest from attendees and the media. Millions of people dream of being able to travel in space, at warp speed, view or visit planets, moons, other star systems and galaxies. Star Trek, Star Wars, countless sci-fi movies, books and magazines have given millions of people a taste of the possibility of space tourism. Consequently, space tourism remains the ultimate travel aspiration. A few wealthy tourists have made commercial visits to the international space station at immense expense, and Virgin has sold tickets and accepted deposits for Virgin Galactic's first commercial 'weightless in space' flights since 2008.

Despite many years of Virgin Galactic test flights, by the end of 2020 the first commercial flight had yet to lift off. The risks involved in space travel for humans are immense and all forms of air and space transport emphasise the safety for passengers and crew. The technology for spacecraft requires multiple backups because there are no mechanical repair shops in space. At the very least, passengers in any space venture would need to show they have the physical capability to survive the rigours of space. A trip on a Virgin Galactic may not be the ideal gift for a frail 90-year-old grandparent but it may suit a super-fit 90-year-old grandparent. Space tourism will occur; it's a matter of when. However, the risk regimes for the complex technology involved in taking passengers into space and returning them home safely take time to develop, as was the case with manned and unmanned scientific space exploration and research.

Case Study 10.1

Cyber Scares for Small to Medium Tourism Enterprises

The global tourism industry is dominated by small to medium enterprises, and one of the challenges faced by small to medium businesses is the cost of computer technology to effectively conduct business, secure and backup the data, and ensure the best protection of computer systems and data against viruses, trojans, hackers and cyber criminals.

A suburban travel agent in most cities and towns, prior to Covid-19, normally had a staff of about five people. However, there is a growing number of one-person travel agents who operate from home using laptop or even a tablet to run a small business. In Australia, there are almost 1,000 such agents operating under the umbrella of the home-based agency networks, travel managers, travel counsellors or mobile travel agencies. Most high street travel offices have computer reservation systems linked to one of the three GDS: Travelport Sabre or Amadeus. This programme controls most bookings and financial transactions relevant to the travel booked. The GDS systems have the capability to manage day-to-day accounting and access to a wealth of travel-related information and news. Many small travel businesses run other

(Continued)

computer systems linked to the internet separate from the GDS network. In some cases, depending on the agency's affiliation with a larger travel agency body it is possible there is either hardware or software to link the agency with its parent company.

Most small businesses and especially travel-related businesses are almost entirely dependent on effective IT systems to manage the business, process bookings, build and maintain client databases, communicate with their stakeholders and customers, and keep up to date with news that impacts on their business.

Skye Theodorou's honours thesis *Cyber Scare* (2017) was also written as a research study for the NSW's Government's Department of Small Business on strategies to assist and train small businesses on computer security against cybercrime. This research was highly relevant to 95 per cent of travel businesses in Australia. The highest rate of cyber threat was access.

The key cyber risks for small businesses identified in the study (in which over 1,000 small business managers in NSW were interviewed) were as follows:

- Interference with, theft and compromising of financial transactions
- Information theft from customer databases
- Identity theft
- Breaches, interference and hacking of confidential correspondence
- Hacking and theft of booking details
- Problematic internet connection
- Credit and debit card scamming
- Fraudulent transactions using stolen ID and credit/debit card details
- Poor backup or lack of data backup
- Hacking of business accounting records
- Crash of a server or crashing of computer hardware
- Inappropriate and outmoded software.

The majority of the risks and threats to small businesses arose from acts of cybercrime. As Theodorou identified, the average cost to a business in both lost business and the cost to correct IT issues was about US$40,000. For small businesses, cyber crime and IT problems have the potential to amount to an existential threat to the business.

Theodorou's survey indicated that over 60 per cent of small business managers surveyed lacked the confidence or knowledge to effectively manage IT problems and cyber threats, and many cut corners (financially) to obtain assistance from cut-price IT 'experts' who lacked the expertise to attend to the IT problems affecting a small business.

As a result of Theodorou's research, the NSW Government's Department of Small Business introduced a range of free online and face-to-face educational and training packages to assist registered small businesses to manage IT issues and cybercrime threats. The courses were designed to enable business managers and selected employees to identify threats and apply cyber-defences and encryption to minimise the likelihood of cybercrime. Travel and tourism businesses in NSW have embraced these programmes. It is well understood that cybercrime is an evolving threat to online security requiring regular update and maintenance.

THE ROLE OF TECHNOLOGY IN TOURISM RISK, CRISIS AND DESTINATION RECOVERY

The one area in which there has been a significant amount of academic research on the technological dimension of tourism risk, crisis and recovery has been in the issue of crisis communication and recovery messaging. The key authors have been mentioned earlier in this chapter. The internet has revolutionised tourism risk, crisis and recovery management. At government level, foreign ministries issuing government travel advisories adopted the internet to maintain, update and disseminate government travel advisories from 2001 (Beirman, 2006 and 2019). Risk management companies are able to conduct online risk audits with their customers.

All sectors of the tourism industry have developed websites to market and promote products, services and destinations, and have used these websites to address a plethora of risk management and safety issues addressed to consumer and business stakeholders. The UN World Tourism Organisation (UNWTO), Pacific Asia Travel Association (PATA) and the World Travel and Tourism Council (WTTC) have been among the many national and transnational tourism associations which have used websites and apps to address risk management issues, especially tourism safety. During the 2020 Covid-19 outbreak, messaging on the status of border closures, quarantine requirements and social distancing has been maintained and updated in real time by national, state and regional tourism authorities.

When businesses or destinations have experienced crisis events such as natural disasters, terrorist attacks, pandemics or crime waves, technology has provided multiple tools to disseminate information on the status of tourism destinations and businesses in real time. Many businesses and destinations have effectively employed social media to counter some of the inaccurate reporting of media outlets. The partnership between PATA and the Tourism Authority of Thailand (TAT) during the 2011 Thai floods exposed fraudulent media coverage, which suggested that Thailand's main international airport was flooded. The airport involved was a disused airport where planes were being disassembled for scrap (Van Walbeek, 2012). Effective use of tech tools such as Google maps have enabled destination authorities to disseminate accurate information and expose false media reporting. The use of websites and social media in the area of fact checking has become increasingly employed in crisis and recovery management, and has extended into the broader business world and political journalism.

Tools such as Trip advisor have been a valuable aid in tourism marketing. It also has value in obtaining real-time consumer information on the status of destinations and businesses perceived to be in crisis. Tackling negative perceptions through the widespread dissemination of factual and objective information via apps, websites and social media is a vital tool for tourism businesses, and in many cases, it has accelerated destination and business recovery when employed ethically and effectively.

Nepal based its recovery from the 2015 earthquake disaster on effective employment of the Nepal tourism Facebook site in which tourists could contribute observations in real time (Beirman et al., 2018). In fact, social media, websites and apps have the benefit of being inexpensive platforms with massive reach. They are excellent tools to disseminate real-time status updates on access, transport links, availability of accommodation, operational status of attractions, shops and entertainment venues.

The wide usage of hand-held devices, tablets and computers has enabled both service providers and destination managers to communicate instantly and accurately with consumers and business-to-business (B to B) stakeholders when a crisis occurs, and at all stages of resolution and recovery. Many of the readily available programmes and apps are inexpensive and the cost of website and app development is low. Consequently, access and usage of social media, websites and apps have been democratised globally, and are widely available to tourism destinations and enterprises in both the developed world and the developing world.

During the 2020 Covid-19 outbreak, video conferencing using platforms such as Zoom and GoToMeeting have become pervasive during a period when social distancing restrictions, lockdowns and travel restrictions have been the norm. In terms of disseminating knowledge, new developments and tourism education programmes using these platforms have been invaluable for tourism businesses, tourism educators, destination marketing/management organisations and global tourism associations. The platforms have enabled the tourism industry to communicate, share and exchange knowledge on managing the pandemic online with thousands of colleagues around the world. Technological advances have been a significant asset to global tourism in recent years.

SUMMARY AND CONCLUSION

The technological dimension is one of the most under-researched aspects of tourism risk, crisis and recovery management. There is an erroneous tendency among many students, scholars and tourism practitioners to view technology from the limited perspective of IT. Important as IT is, this chapter has demonstrated that this is just part of a far broader context of technology and the risks associated with technology.

Tourism has a high level of dependence on technology when it comes to tourism-related products and service delivery. Advanced technology is embedded in all tourism-related transport, accommodation, event venues, attractions and especially in the booking and confirmation of tourism services. All of these technologies are accompanied by attendant risks.

As most tourism businesses are small and increasingly dependent on IT, the protection of IT technology from breakdowns and cybercrime is a top priority for all tourism businesses. Backup, either physical or in the cloud, is vital to preserve sensitive data relevant to the business.

Advances in IT technology during the 21st century have proven to create a vast array of tools, which has aided the visibility, dissemination and timeliness of risk, crisis and recovery communication. Websites, apps and social media have the ability to disseminate verifiable facts, counter false reporting and challenge negative and inaccurate perceptions. These technologies and platforms are affordable to large and small businesses and the destination management organisations of rich and poor nations alike. IT advances since 2000 have democratised tourism risk, crisis and recovery management in the same way the wide-bodied, high-capacity jet led to a reduction in air fares for many years after the Boeing 747 jumbo jet came into service in 1970.

Discussion Questions and Issues

1. Transport technology has a vast range of attendant risks. Choosing one mode of transport, discuss the key risks associated with this and the main treatments of those risks.
2. You are managing a destination of your choice, which has experienced a natural disaster of your choice. What would be your IT approach to communicating your management of the crisis and the recovery process?
3. You are running a small tourism business. What measures would you take to minimise the exposure of your business to cybercrime and data theft?
4. You are managing a restaurant, which received a terrible (and in your opinion grossly unfair) review in the media. Utilising IT, how can you address the negative reviews and restore the previously good reputation of your restaurant?

REFERENCES

Amadeus (2020) *Amadeus Global Distribution System*. www.amadeus.com. Accessed 9 December 2020.

Beirman, D. (2006) 'A travel industry perspective on government travel advisories', in Wilks, J., Pendergast, D. and Leggat, P. (eds), *Tourism in Turbulent Times. Towards Safe Experiences for Visitors*. Elsevier Amsterdam. pp. 309–22.

Beirman, D. (2012) 'The impact of volcanic ash clouds in 2010 and 2011 on global tourism mobility and international transport', Best Education Network, Think Tank, 12 June 2012, France.

Beirman, D. (2019) 'Collaborative approaches to government travel advisories in Australia between Australia's tourism industry leadership and the Australian Department of Foreign Affairs and Trade'. *Journal of Vacation Marketing*, 25(1): 71–87.

Beirman, D., Upadhayaya, P., Pradhananga, P. and Darcy, S. (2018) 'Nepal earthquake in the aftermath of the April/May 2015 earthquake and aftershocks. Repercussion, recovery and the rise of new tourism sectors', *Tourism Recreation Research*, 43(4): 1–11.

Cordato, A. (2017) 'Dreamworld tragedy highlights legal duty of care for theme park guests', *Australian Leisure Managers*. https://ausleisure.com.au/news/dreamworld-tragedy-highlights-legal-duty-of-care-for-theme-park-guests. Accessed 18 December 2020.

Cruise Lines International Association (CLIA) (2020) https://cruising.org/en. Accessed 2 December 2020.

Gennawey, S. (2013) *The Disneyland Story. The Unofficial Guide to the Evolution of Walt Disney's Dream*. Unofficial Guides.

Henderson, J. (2007) *Tourism Crises, Causes, Consequences & Management*. Elsevier: Burlington, MA, USA. pp. 124–40.

Hern, A. (2016) 'Using Wifi in Air BnB rentals poses security threats', *The Guardian*, 6 August 2016.

International Maritime Organisation (2020) www.imo.org. Accessed 2 December 2020.

Merlin Entertainments (2020) www.merlinentertainments.biz. Accessed 4 December 2020.

Park, D., Kim, W.G. and Choi, S. (2019) 'Applications of social media analytics in tourism crisis communications', *Current Issues in Tourism*, 22(15): 181–5.

Pennington-Gray, C., Schroeder, A., Donahoe, H. and Klol, S. (2013) 'Using social media in teams', *Journal of Travel and Tourism Marketing*, 30(1–2): 126–43.

Porter, M.A (2018) A size comparison of the Titanic to Modern Cruise Ships. https://owlcation.com/A-Size-Comparison-of-the-Titanic-to-Modern-Cruise-Ships. Maven Publishing USA. Accessed 12 June 2021.

Sigala, M. (2012) 'The impact of geocollaborative portals on group decision making for trip planning', *European Journal of Information Systems*, 21(4): 404–21.

Theodorou, S. (2017) *Cyberscare. A Study on Small to Medium Sized Enterprises and the Emergence of Cyber Security Risks in Australia*. Bachelor of Management, University of Technology Sydney, Honours Thesis, June 2017.

Van Walbeek, B. (2012) *Crisis Management Presentation*. PATA. www.pata.org. Accessed 9 December 2020.

Whittaker, Z. (2019) 'Flaws in Amadeus airline booking systems made it easy to change passenger records', *Tech Crunch*, 16 January 2019.

Young, A. (2019) National Sales Manager Contiki Holidays (Australia), Personal Communication. Guest lecture to University of Technology students at the Sydney HQ of the Travel Corporation, 16 April 2019.

11

THEME 8: RECOVERY FROM INTERNAL MANAGEMENT, PRODUCT AND SERVICE FAILURES

INTRODUCTION

Most of the risks and crises discussed in this book and academic tourism, risk, crisis and recovery literature arise from events and circumstances, external to and beyond the direct control of destinations and enterprises. However, a vast number of crises, especially those that compromise the reputation or viability of an individual business, industry sector or a destination arise from failures in management at both the staff management level and financial management level or shortcomings in product, service or customer relations. Many tourism and hospitality companies have been accused of over-promising and under-delivering in the product and service chain. In an increasingly litigious world accompanied by increased consumer power through online social media and consumer sentiment sites such as Tripadvisor, consumers have never been more empowered to hold tourism businesses to account when their tourism experience fails to meet the expectations promoted and marketed by tourism businesses and destinations.

This chapter will look at the broad issue of management, service and product failures and focus on a number of significant cases, which illustrate specific aspects of these failures. In some cases, these failures led to the demise of a company, and in others a downturn was followed by a recovery. The three main cases to illustrate these points are the Thomas Cook Collapse (BBC, 2019), United Airlines service problems (2017) and Costa Cruises' mishandling of the *Costa Concordia* (BBC, 2015) and *Costa Allegra* mishaps (BBC, 2012).

Tourism, in common with all service businesses, has experienced product and service failures. Usually, the response to complaints about poor service or sub-standard product involved effective and dedicated complaints handling actions and processes. In some cases, issues related to the complaint were exacerbated by poor customer complaints handling attitudes and processes. These include:

- Multiple stages of managing a complaint. This sometimes occurs when a complaint to a travel retailer involves an external party in the tourism service chain.
- Complex bureaucratic processes involving the complainant being required to provide extensive and time-consuming documentation.
- Response time by the company far later than initially promised.
- Failure by companies to return complainant phone calls or correspondence.
- Unreasonably lengthy response times as part of a company's complaint handling policy.
- Attitudes to customers on the part of companies, which led customers to perceive that their complaint was treated as an inconvenience, false or invalid.
- A reluctance or refusal to compensate or apologise for service failure.

Appalling customer relations make for great comedy. The enduring popularity of the 1970s British comedy series *Fawlty Towers* was based on viewers identifying with the experience of encountering crass 'hospitality manager' characters such as guest house owner Basil Fawlty. Basil's philosophy was, 'the customer is an inconvenience and is nearly always wrong'. This is fine in the world of comedy but is a self-defeating attitude in the real world of business and especially in tourism businesses so heavily based on positive experiences. In essence, product and service failures are largely own goals in business and depending on how they are managed or mismanaged they can serve to either enhance the business or destroy it.

Although tourism growth until 2019 was largely facilitated by technological development, in essence, tourism remains a people-to-people business. Ideally, tourism and especially leisure tourism is largely about the fulfilment of aspirational experiences for people ('the dream factory'). When a paying customer's product and service falls short of those aspirations, we see the first symptoms of management failure. Many consumers have a variable degree of tolerance for errors or problems, which arise during their tourism experience. The more elements included in a complex tourism programme the greater the likelihood for something to go wrong. An unannounced change in the departure time of an inter-island ferry; sub-standard hotel rooms; rude, impatient or incompetent tour guides; tour coach breakdowns; unscheduled closure of an attraction; incorrect documentation details of hotel bookings, airline flights or meal plans on flights and tours; and inaccurate tour or restaurant bookings – are among many frustrations that tourists experience during the course of a holiday trip. While travellers may laugh about these mishaps later, at the time, they detract from their travel experience.

Travel agents and wholesale tour operators generally undertake exhaustive checking processes to minimize errors along the service chain. This is an integral part of a tourism business' commitment to *duty of care* (Cordato, 2017). Failure on the part of the travel agent or tour operator to practice due diligence in matters of documentation and fulfilment of marketed product and services can result in dissatisfied customers who can utilise substantial personal and social media networks to communicate their dissatisfaction. This could potentially damage the company's reputation. In some cases, these service failures could lead to litigation. Tourism businesses should be well aware of their legal obligations to their clients. Most legal jurisdictions have strict laws regarding duty of care, transparent and inclusive pricing, honest marketing and safe business practices. In litigious societies (which includes most Western countries), tourism law has been a growing sector within legal firms and practices.

For most travel arrangements, organised by a travel retailer or wholesaler, the service chain starts with travel advice and filtering the travellers' key wants, needs and budget. This will also include advice on formalities such as visa requirements and vaccinations (when appropriate). The next step involves the mode of transport, route and carrier. The third step normally involves meet and greet at the destination gateway point followed by a transfer to the initial night of accommodation. The tour programme will involve an agreed form and standard of accommodation, tour transport, a high-quality tour guide, included meals, visits to attractions and special tour highlights. The documentation provided to the traveller, including details of accommodation,

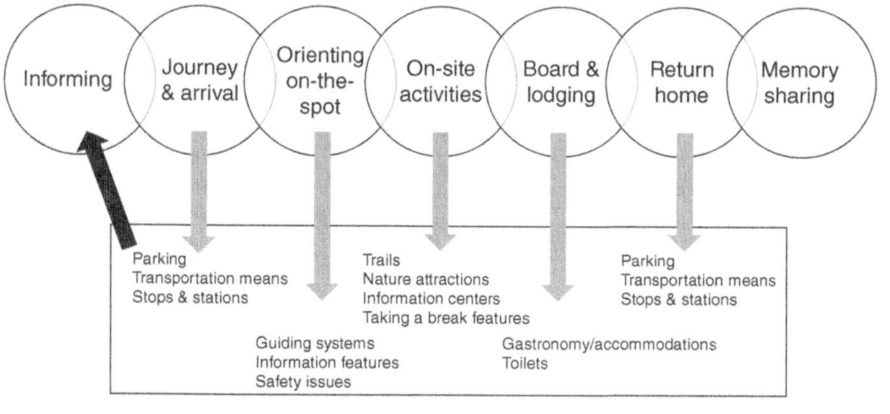

Figure 11.1 The tourism service chain as it applies to complex eco-tourism arrangements

Source: Hennig et al. 2015. Used with kind permission of Sabine Hennig.

ground operators and itineraries, are expected to be correct. However, due to the many links in a service chain, even with the best companies, errors do occur. The greatest challenge for travel agents and wholesale tour operators in packaging a tour programme involves ensuring that the multiple elements that make up a tour programme meet the level of quality and reliability to fulfil the marketing claims of the tour operator. The diagram in Figure 11.1 provides one of many examples of a tourism service chain. Although problems and errors at each link in the chain may not necessarily be the fault of the travel agent or tour operator, from the customer's perspective the person or company they booked with is ultimately responsible, and in most cases the only point of contact they have to resolve service problems.

Many companies measure the effectiveness of their service delivery through a model called servqual (service quality). Servqual assesses the quality of business-to-business (B to B) and business-to-consumer (B to C) service throughout the service delivery chain. As Figure 11.2 illustrates, the core elements of good service in tourism involve:

1. Reliability of maintaining a high level of quality customer service.
2. Tangibility that the tour products are as promised in the brochure and are matched by the quality of the coach, hotel room, airline flight, included meals and attractions etc.
3. Assurance that intangible service promises (including attitude to clients and the quality of the tour experience, sufficient time to visit attractions) are met.
4. Empathy towards the special needs of the customer. Empathy also encompasses a positive attitude towards and respect for the individual needs of customers.
5. Responsiveness to customer needs and problems associated with the trips when and if they arise.

Today, due to the enhancements of consumer powers and rights and consumer access to social media, most tourism businesses are aware that their reputation as a tourism service provider is subject to rigorous assessment by consumers and business stakeholders along the service chain.

One of the many popular YouTube videos on poor service response involved a performance from a country group led by Dave Carroll (2009) who composed a catchy song called 'United Breaks Guitars', which related a true story about baggage handlers from United Airlines in 2009. United Airlines baggage handlers damaged a valuable guitar belonging to one of the band members while travelling with the airline within the USA. After the band lodged a complaint and a request for compensation, United Airlines refused to take any action. 'United Breaks Guitars' became a smash hit for the group and the YouTube version of the song has been viewed by over 20 million people. So damaging was this incident to United's reputation that its 2009 share price and company value dropped by ten per cent (or US$180 million). The airline finally relented and compensated the group for the damage to the guitar. A very expensive service failure! Ironically, had United Airlines compensated the group, there would have been no 'United Breaks Guitars' smash hit and United's share price would not have declined during 2009. As we will see later in this chapter, United Airlines did not learn its 2009 lessons of customer relations.

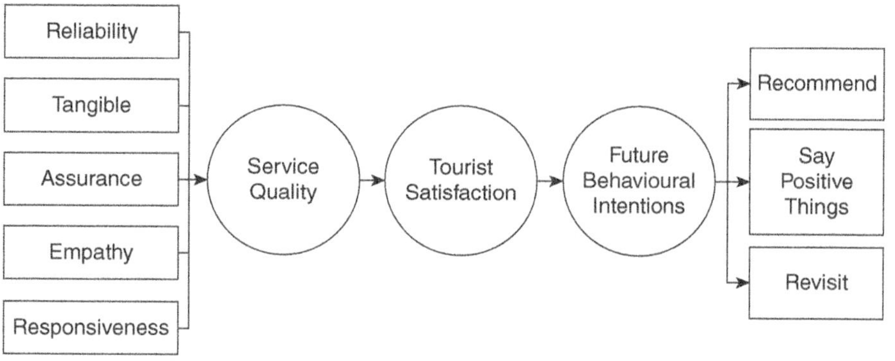

Figure 11.2 Core concepts of servqual in tourism consumer orientation

Source: Figure courtesy of Canny, 'The role of food quality service and physical intertemporal consumer care environment on customer satisfaction and future behavioural intentions in casual dining restaurant', Microeconomics: Intertemporal Consumer Choice and Savings, E Journal. Copyright Elsevier (2013). Used with permission.

One of the challenges faced by service providers involves responding appropriately when service failures become public knowledge through negative media (including social media) coverage. The tabloid media, across all its platforms, thrives on covering stories of unsatisfactory service, overpricing and human-interest stories of customers who lost their life savings due to travel deals with dubious companies, or when their travel company collapses. Critical to dealing with all crisis events is an ability to engage appropriately with the media and in fact all relevant stakeholders. When a company's reputation is at stake, the ability to communicate effectively and honestly to all stakeholders, including the media, is often the difference between business survival and extinction.

Tourism businesses are frequently vulnerable to corporate collapses. As discussed in Chapter 3, the prevailing business operation model of high volume and low yields, prevalent in all sectors of the tourism industry, has proven to be a source of risk and vulnerability, especially when travel businesses are subject to disruption by external crises, including Covid-19. Low yields result in travel businesses that lack financial reserves enabling them to survive prolonged business downturns. During the Covid-19 crisis in 2020, there has been a virtual conga line of travel businesses (begging bowls in hand) seeking government support for their sector. In August 2020, STA Travel (a very popular travel agency among students) was one of many high-profile, multinational tourism industry companies that collapsed due to the lack of financial reserves to ride out a challenging business environment. However, many more companies have collapsed due to poor management decisions.

Case Study 11.1

The Thomas Cook Collapse, September 2019

Globally, the most prominent travel industry collapse in recent years was the demise of Thomas Cook in September 2019. The Thomas Cook collapse essentially involved an iconic, 178-year-old tourism brand, overreaching itself and placing itself in such a high level of debt that it was beyond rescue.

The concept of multi-branding for many tourism businesses has been a successful strategy to appeal to various segments of the tourism market. For most travel companies, multi-branding has been a function of horizontal integration, different brands providing a similar service pitched at different market segments and price points. The Travel Corporation operates over 25 brands of tours pitched at very specific segments. Many of its brands, including Contiki Tours for 18s–35s, Insight Vacations, Trafalgar Tours and Busabout, are better known than the parent company. Some airlines have successfully multi-branded. Singapore Airlines, Scoot, Tigerair and SilkAir are all divisions of Singapore Airlines, appealing to specific categories and price points. The long-term success of the Travel Corporation and Singapore Airlines is that both companies played to their strengths as either a tour operator or an airline.

Most tourism students are familiar with Thomas Cook as the pioneer of the package tour and the travellers' cheque. Cook's first business venture was in 1841 when he organised one-day rail excursions between Leicester in central England to the neighbouring town of Loughborough.

By 1855, Thomas Cook organised tours on the European continent, which included transport, accommodation, meals, guided visits to attractions and special events. This was the forerunner of the international package tour. In 1865, he opened his first high street travel agency in London, which was the retail arm enabling him to sell his growing range of domestic and international package tours. By 1874, Cook issued traveller's cheques (initially from the USA) exchangeable for cash in many countries (Weaver and Lawton, 2014).

Thomas Cook experienced many changes in ownership (including a period between 1948 and 1972, when it was state owned and managed by the British Transport Holding Company), but it chose the path of vertical integration. It sought to provide services along the vertical tourism service chain. This included travel agency services, package tours, accommodation, financial services, insurance, coaches and by 2007, an airline that was part of the merger between Thomas Cook and MyTravel Group. The fundamental problem with this approach is that some links in Thomas Cook's vertical integration chain were stronger, more profitable and better managed than some other divisions (BBC, 2019).

The airline division was especially problematic. Although Thomas Cook had a substantial package tour market for its airline, the airline was an ongoing loss-maker. By February 2019, Cook had received bids for its airline business. Had it accepted the bid from Condor, Cook's level of debt may have been reduced to manageable proportions.

Thomas Cook also diversified into a hotel booking website in 2008 (Hotels4U) and a media group (Octopus Media Technology) to host Thomas Cook TV. One of Thomas Cook's major loss-making divisions was its extensive network of 600 travel agencies in the UK, Europe and the Middle East.

The writing was on the wall for Thomas Cook when in May 2019 it announced a loss of £1.5 billion for the first half of the financial year. In June 2019, the Chinese company, Fosun International, offered to pay £450 million for 75 per cent of Thomas Cook's tour business and 25 per cent of its airline. Despite Fosun's offer to add another £200 million to its original investment, the British government decided not to assist Thomas Cook financially. On 23 September 2019 Thomas Cook was placed into administration with virtually no advance notice and ceased trading. This led to an immediate loss of 21,000 jobs from CEO to the most junior staff members, the immediate grounding of its aircraft and the closure of 600 agency offices. The collapse also left over 600,000 British and EU travellers (Thomas Cook customers) stranded. Thomas Cook's collapse was not due to a lack of turnover. Thomas Cooks' various services attracted a vast number of customers but the yield per passenger was so low that the level of turnover failed to make inroads in reducing Thomas Cook's debt. This was clearly identified by financial institutions but not by Thomas Cook's management, which incorrectly assumed that the company's name and historic reputation would be sufficient collateral to support the debt.

The Thomas Cook collapse would have been an unmitigated disaster for its 600,000 stranded customers but for a superb example of risk management instituted by both the UK government and the European Union (EU). Unfortunately, the management and staff of Thomas Cook, most of whom were unaware of the company's financial travails, lacked the protection of any compensation package.

The risk of corporate collapse in tourism is a significant risk globally. As discussed in Chapter 3, the predominant business model of tourism is high volume and low yield. This means most tourism businesses across all sectors operate on very slim profit margins and depend on high turnover to make a profit. Many travel companies fail to achieve both and consequently when the business environment is difficult, they lose money.

In 1973, the British government supported and backed the establishment of the Air Travel Organiser's Licence (ATOL, 2020). This scheme was a consumer protection scheme designed to cover consumer losses in the event that a travel company became insolvent. Effectively, ATOL required that package tour companies (usually paid for as a small surcharge by consumers) pay an amount of £2.50 per passenger/per booking for coverage of travellers' costs, including repatriation to their home port in the event that their travel company became insolvent. The scheme was government backed and only applied to package tourism (a combination of air and land content). In 1990, the EU commenced a similar scheme (updated in 2001) known as the European Package Tour Directive (2020).

The funds, which were built up by consumer or company payments into the schemes over many years, were sufficient to cover the repatriation and refund costs of 150,000

(Continued)

UK and 450,000 European tourists whose trips were subject to either cancellation or mid-trip disruption.

The Thomas Cook collapse, while a disaster for the company and its employees, was an outstanding example of forward-thinking risk management policy to protect tourism consumers of package tour products. A sequel to the Thomas Cook collapse involved the Chinese company Fosun International, re-launching a slimmed down version of Thomas Cook in July 2020. It can only be hoped that the new iteration of Thomas Cook avoids the over-diversification of the company and the accumulation of debt that led to the demise of the old Thomas Cook.

The Thomas Cook case is an instructive case for tourism businesses which over-extend themselves into sectors of tourism in which they have little expertise, human and financial resources or knowledge. From a risk management perspective, the decision to engage in vertical integration meant that very different risk management protocols and policy frameworks would apply to each sector on the product chain, including airlines, accommodation, travel agents, tour operators and media services. This greatly complicated risk management for Thomas Cook, which was already coping with underperforming parts of the business.

MAKING MULTI-BRANDING WORK FROM A RISK MANAGEMENT PERSPECTIVE

In the global tourism industry, the most successful multi-branded companies are those which focus on primarily horizontal integration. The main reason for this is that the brands are essentially providing a comparable service, although pitching this at different markets. Companies such as Carnival Cruising (cruising), the Travel Corporation (tour operators and accommodation providers), Singapore Airlines (aviation), Merlin Entertainments and Disney Corporation (attractions) and Accor and Intercontinental Hotel Group (hotels) have been successful multi-branded companies largely due to the fact that they focus on one or two sectors of tourism and the risk management expertise can be applied to all its brands.

From a risk management perspective, horizontal integration enables risk management policies and practices to apply to all brands within the corporate group, especially if they are operating in the same sector. As an example, Merlin Entertainments (2020) has 150 brands, all of them attractions, but a single set of risk management and safety protocols applies to all the brands. This also applies to the hotel sector where all hotels (large, small, five star or three star) face a wide range of common risks.

By contrast, multi-branded companies that practice vertical integration have faced the challenge of needing to apply different sets of risk management skills to various brands.

The essence of the Thomas Cook collapse discussed earlier in this chapter was that risk management practices for airlines, financial services, insurance, hotels, tour operators and travel agents are very different and require extraordinary skills to effectively coordinate them all under the same corporate framework.

The Travel Corporation (2020) is the parent company of the world's largest multi-branded tour operator and is essentially a family run company, run by the Tollman family. Many of its 40 brands are household names in global tourism, including Contiki Holidays for 18s–35s, Insight Vacations, Trafalgar Tours, Busabout and Thompson Holidays. With the notable exception of the Red Carnation Hotel Group, nearly all Travel Corporation brands are tour operators, river boat cruises or tour coach companies. Each brand division runs its own operation and markets and sells its own distinct products and services pitched to a specific sector of the travel market. However, the Travel Corporation pools its resources in terms of HR management, risk and crisis management, reservation systems, negotiation with partner principals, notably airlines, accommodation providers, attractions and restaurants.

Risk management for tour operators is quite complex because the tour operator is rarely the sole provider of the tour package. The Travel Corporation companies tend to employ their own guides, own their own vehicles or vessels and for selected upmarket tours which use Red Carnation hotels, they occasionally own the accommodation. However, they don't own restaurants, airlines, the majority of hotels they include in their programmes, cruise ships, on-tour attractions and some specialist tour components. The Travel Corporation applies stringent risk management protocols for their tour programmes, especially those which involve adventure tour elements, quite a common element in a Contiki Tour. Those services contracted out by Travel Corporation brands are required to meet the Travel Corporation's strict risk management requirements.

The Travel Corporation approach to risk and crisis management faced its sternest test as a result of a disaster in 1999. A group of mainly Australian, British, South African and American travellers on a Contiki tour in Switzerland undertook a Swiss Canyon hike operated by a contracted Swiss tour company, Adventure World. As a result of a flash flood that occurred during the hike, 21 people (three guides and 18 young tourists of whom 14 were Australians) died from being swept away by the raging torrent in a narrow canyon. The disaster sent shock waves through Contiki Tours, along with all the Travel Corporation brands. Its long-term impact was to heighten the entire company's risk management awareness and strengthen its protocols, especially for its adventure tour programmes. The Swiss canyoning disaster also led to the Travel Corporation's brands to adopt far stricter checking procedures for contractors.

One of the greatest challenges for any travel company faced with a disaster in which its clients are killed in the course of a tour programme is to rebuild confidence and trust within the organisation and all its stakeholders, including customers.

Peter Smith (2018), CEO of Contiki Australia at the time of the disaster, took it upon himself to maintain contact with the families and friends of all the Australian victims and the survivors, and to be the point of contact with the media and all other stakeholders. Contiki Australia arranged for Australian families of victims to visit the site of the disaster when they chose to. Although the Swiss canyon tour was run by a contracted operator, Contiki and the Travel Corporation (as the wholesale operator) assumed full responsibility and accountability for the disaster. While there was damage to the brand's reputation in the immediate aftermath of the tragedy, Contiki's compassionate, responsible and professional response to the disaster did eventually restore both the Contiki brand's and the Travel Corporation's reputation in the long run. As a parent company, the Travel Corporation was careful to avoid collateral damage to the reputation of its other brands, and it worked very closely with Contiki's management to address reputational damage in its dealings with all relevant stakeholders.

Case Study 11.2

United Airline Turns 'Re-Accommodation' into a Term of Derision and a Reputational Crisis

Most of the world's airlines have engaged in the practice of overbooking. An airline has a finite number of seats available for each flight. On most flights, a number of passengers either cancel their flight at the last minute or simply don't make the flight for any one of a number of reasons. In such cases, no-show passengers or late cancellations are charged a cancellation or no-show fee, which is usually the equivalent of the fare paid. For passengers who have taken out appropriate travel insurance and have compelling reasons for a no-show or cancellation, cancellation fees are covered by travel insurance. In most instances, the airline is financially compensated when passengers fail to utilise their seats. In many cases, last minute or pre-paid standby passengers are able to purchase or fill vacant seats. This actually means the airline gains revenue above the level of a full flight.

Aviation, in common with most sectors of the travel industry, is a highly perishable business. A high percentage passenger load is integral to the profitable running of an airline. In view of the high statistical likelihood of no-shows (Laurence et al., 2003), which is about five per cent of passengers, airlines have engaged in the practice of overbooking airline seats for many years. In the vast majority of cases, overbooked passengers are boarded to replace no-shows. On those rare occasions in which all passengers attend the flight, those passengers who were the last ones booked are usually asked to postpone their flight, and under IATA rules (2017), are entitled to either monetary or in-kind compensation from the airline, which may range from a meal to overnight accommodation depending on the location and duration of the delay. This applies to the 85 per cent of the world's airlines that are IATA members. Many low-cost carriers are not IATA members and some low-cost carriers refuse to compensate affected passengers. In most cases of this type, passengers are asked to volunteer to be excluded (bumped) from a flight in return for the compensation the airline offers.

The US-based United Airlines is one of the world's largest and oldest commercial carriers. The airline commenced operations in 1928. In 2010, United absorbed another large US carrier, Continental Airlines. United has survived a contraction in the number of viable US-based commercial airlines since 2000. Famous brand names, including Pan Am, TWA, Braniff and Northwest Airlines have disappeared but United has survived. In 2019, United Airlines operated an extensive network of 242 domestic US destinations and an extensive network of 120 international routes in 61 countries. In 2019, it flew 162 million passengers and had a fleet of 788 passenger jets (United Airlines, 2019).

Given the vastness of these figures, it comes as little surprise that not every United Airlines passenger is 100 per cent happy with its service. Its company slogan 'Fly the friendly skies' has been the airline's slogan for many years and there have been occasions in which the airline's service has fallen short of its marketing slogan. Earlier in this chapter, mention was made of the 2009 hit song 'United Breaks Guitars'. The sheer volume of passengers and the variety of service expectations are a challenge to any airline. In common with any airline with such an extensive operation and long history, United has experienced its share of crises.

Arguably, the most serious crisis United Airlines ever experienced was on 11 September 2001 when two United Airlines planes were among the four US airlines hijacked on that fateful day. UA 175 was flown into the south tower of the World Trade Center in New York City. UA 193 was also hijacked, following a fight between the hijackers and passengers for control of the aircraft, the airline crashed in a field in Somerset, Pennsylvania, killing all passengers and crew. The impact of 9/11 affected the entire global airline industry and led to a global review of safety and security for airlines and airports. As United and its competitor American Airlines accounted for all four of the aircraft hijacked on that fateful day for aviation, the International Civil Aviation Organization's (ICAO) approach to airline recovery treated the terrorist attacks as a systemic crisis for aviation security rather than as security failures on the part of either United or American Airlines.

United Airlines' 2017 overbooking crisis killed nobody but caused significant damage to the reputation of the carrier. On Sunday 9 April 2017, United Airlines Flight 3411 was flying a US domestic route from Chicago to Louisville. The airline wanted to substitute four crew members for paying passengers who were seated, booked and ready for departure. Three paying passengers accepted United's compensatory travel

(Continued)

vouchers and left the flight without incident but no other paying passenger volunteered to disembark. United then asked airport security to forcibly remove a passenger, David Dao, a Louisville-based medical practitioner of Vietnamese origin, from the flight. The YouTube video of the violent removal of Dr Dao from the flight went viral, especially in China where over 100 million people (most of whom incorrectly thought this action was targeted at a Chinese man) watched the video in a single day. United's action was roundly condemned in China.

Dr Dao suffered a broken nose, two lost teeth and concussion in the course of his forcible removal from UA3411. If the forcible, bloody and noisy removal of a screaming Dr Dao wasn't bad enough, the additional dimension of racism, whether intentional or perceived, was rapidly turning into a PR disaster for United Airlines (Benoit, 2018).

United's CEO, Oscar Munoz apologised for the inconvenience experienced by all passengers involved in the flight with the notable exception of David Dao. He suggested that Dr Dao's belligerence caused the incident. In attempting to explain United's actions, Mr Munoz stated that United was upset by the event. United made a public apology for having to 're-accommodate these customers' (Levin, 2017). The term 're-accommodate' would long live in infamy as the slogan for appalling customer relations (Benoit, 2018).

It took a significant level of negative consumer sentiment and a massive drop in United's share price to ram home the message to United Airlines that their handling of this incident was a monumental disaster. A Harris poll of 2016 indicated that seven per cent of United's passengers thought United had a bad or very bad reputation. By 2017, the poll taken after the incident indicated that this opinion was now held by 42 per cent. A survey conducted by Public Policy Polling in 2017, after the 're-accommodation' incident, indicated that 40 per cent of those surveyed thought United Airlines was the worst airline in the US. Two days after the incident, the value of United's shares declined by US$1.4 billion (Shen, 2017).

Oscar Munoz was ultimately persuaded that his initial response was far from perfect. Two days after the incident, he declared, 'it's never too late to do the right thing'. On 11 April he did apologise to Dr Dao (as the 'customer forcibly removed') but did not refer to him by name. Munoz did admit, upon reflection, that United treated Dr Dao badly and on ABC's American TV programme *Good Morning America* on 12 April 2017, Munoz declared that he felt shame and such an incident should never happen again. On this occasion, he referred to Dr Dao by name and promised financial compensation to all passengers on the affected flight. Benoit (2018) described Munoz's delayed response as one of 'mortification'. In other words, he realised his initial response on behalf of United Airlines was harmful to the carrier's reputation and sought to convince all stakeholders (especially customers) that the Airline, management and staff, were truly sorry for this incident.

It is worth noting that Munoz maintained his position as United's CEO well into 2020. While most companies would have summarily dismissed a CEO for an incident that had such a negative impact on the company's reputation, Munoz was retained. This speaks volumes about the corporate culture of United Airlines, which contradicts its market slogan, 'Fly the friendly skies with United'. In addition, United's action nearly created an 'international incident' between the USA and China. United Airlines appears to harbour a corporate culture which has a high level of tolerance for 'mistakes' in customer relations. In the US, the dominance of three major airlines, United Airlines, American Airlines and Delta Airlines in the US domestic aviation market has reached a stage where these three carriers are too big to fail and cannot readily be substituted by smaller carriers with limited route networks.

United Airlines survived the odium of this incident and from 2017 it sought to offer a far more personalised level of service on both international and domestic routes. The issue that led to the incident in this case study – overbooking – continues but United Airlines passengers may feel a modicum of reassurance that should they be asked to 're-accommodate', the request is unlikely to be accompanied by assault causing grievous bodily harm.

MONITORING BUSINESS-TO-CONSUMER AND BUSINESS-TO-BUSINESS CUSTOMER SATISFACTION

Tourism businesses and destination managers need to engage on a regular basis with customers both at the business-to-business (B to B) level and the business-to-consumer (B to C) level to ensure that customers are receiving the speed, accuracy and quality of service that is promised in the marketing message (Weaver and Lawton, 2014). Other significant factors in assessing service quality include attitude, monitoring of the competition, consistency and responsiveness to complaints or problems.

Trip Advisor is one of many online programmes enabling customers to assess a wide range of tourism and hospitality services, including flights, accommodation, restaurants, attractions, cruises and tours. While online feedback is a helpful guide, companies and destinations should undertake a mixture of assessment methodologies to ascertain reliable feedback. In essence, this is basic market research. Focus groups, depth interviews and administered questionnaires are all legitimate means to receive feedback from customers, suppliers and those companies who assist a company to on-sell and distribute products and services.

Many hotels leave a printed survey next to the bed but these are rarely filled out by hotel guests. Response rates to surveys are far greater when they are politely and quickly administered as a customer is checking out. Some customers in a hurry may refuse but most customers value the opportunity to give an opinion of their experience, provided it's not time consuming.

The regular monitoring of customer and stakeholder feedback is not just a market research function, it is also a vital risk management and crisis prevention function. The early identification of service problems with customers or with stakeholders provides a valuable opportunity to address minor problems before they become major problems. A tour operator reservation officer who is rude to travel agents could lead the travel agent to make future bookings with a competitor (Page, 2015).

Quite often, what appears to be a minor irritant can have a severe impact on B to B and B to C relations. The most valuable assets for any tourism business are the quality of their staff, and maintaining harmonious relations and a good reputation with every significant other that the company does business with, especially customers.

Case Study 11.3

Costa Cruises' Annus Horribilus, 2012

Costa Cruises is a branded division of the multi-branded Carnival Cruising Company, one of the world's largest cruise companies. Readers will be familiar with the severe problems that the global cruising industry has experienced during the Covid-19 pandemic of 2020, discussed in Chapter 3. However, apart from the 2020 issues that have affected all cruise companies, Costa experienced two disasters, which were due to failures of the company itself and the response to which fell well short of an ideal crisis management response.

On Friday 13 January 2012, Captain Francesco Schettino was in command of sailing the *Costa Concordia* with 4,200 passengers near the coast of the small Italian island of Giglio. He wanted the passengers and especially two young ladies he was trying to impress to have a closer view of the island (Liu-Lastres and Johnson, 2019). The result of this navigational innovation was that the *Costa Corcordia* struck uncharted rocks, ran partially aground and tilted dangerously on its side. In the panic and confusion that followed, 32 people died, trapped in the ship's lower waterlogged decks. To add to the Captain's navigational blunder, he chose to abandon the ship (he claimed he was seeking help), escape and hide. Although the majority of the 4,200 passengers were traumatised by the 'black Friday' incident, most escaped injury (BBC, 2015).

Captain Schettino was ultimately (in 2015) tried and sentenced to 16 years, imprisonment, bemoaning his fate as a scapegoat (BBC, 2016). The Costa management's response to this disaster, especially on television, was wooden, formal and appeared to have the sincerity of a mass murderer regretting that the noise of his machine gun kept the neighbours awake. Costa ultimately did pay compensation to passengers and the families of those killed but its reputation was severely damaged, and the apparent insincerity of its apologies gained it no friends.

Several weeks after the *Costa Concordia* disaster, Costa Cruises experienced its second own goal. The *Costa Allegra*, an 'ultra-luxury' vessel, was en route from Northern Madagascar to the Seychelles island off the East African coast. The ship carried 600 passengers and 400 crew. On 27 February 2012, a massive fire erupted onboard in the operational cabin, which shut down all onboard and engine power, leaving the ship floating about 200 nautical miles south of the Seychelles. Fortunately, there were no casualties but what began as a super-luxury cruise, rapidly morphed into a very hot 'journey from hell' in the heat of a tropical southern hemisphere summer. Much of the onboard food spoiled and air-conditioned cabins were oven hot. The normally pampered, predominantly elderly Americans and Europeans endured steerage conditions and sandwiches for three days in the glare of global media coverage.

Costa was assailed with law suits from the time passengers left the ship and returned to their respective homes. Costa's senior executives did a reprise of their wooden media performance after the *Costa Concordia*. Although no fault was found with the crew of the *Costa Allegra*, the complete breakdown of the ship's operating and electrical systems sent alarm bells through the Costa Brand and the Carnival Cruise Line parent company.

Although the official announcement in June 2012 of Costa's CEO, Pier Luigi Foschi (stepping down from his CEO position because he had reached age 65 and becoming Chairman of Costa Cruising), bears some of the hallmarks of being 'kicked upstairs', without casting aspersions on his character or his ability to manage a cruise company, his abilities as a crisis communicator proved to be a liability for his company at a time when Costa's reputation was in severe peril.

SUMMARY AND CONCLUSION

Management and service failures or 'own goals' are a frequent cause of crises that lead to the collapse of tourism businesses or severe reputational damage. Tourism businesses and destinations operate in a highly competitive business environment. Service and operational failures on the part of one company present opportunities for competitors to seek custom and loyalty of dissatisfied customers. In common with most service industries, the reputation of a company is a fragile proposition and you are only as good as your latest transaction.

The three case studies illustrate three different types of management failure. The Thomas Cook collapse was the culmination of many years of financial mismanagement, poor management of unprofitable divisions, and a gross failure to consult and communicate with employees and stakeholders. It was financial and systemic failure on an epic scale, which cost 22,000 people their jobs and disrupted the holidays of over 600,000 Thomas Cook customers. The only saving grace was that the customers were rescued by government-backed insolvency protection schemes, which were established due to the fact that Thomas Cook was not unique in collapsing from systemic financial and operational failures.

The United Airlines case was a product of a corporate culture in which corporate needs overrode those of its customers. The fact that United's CEO retained his job years after an incident which damaged United's reputation, led to slump in its share price and lost it many customers demonstrated a culture with a high tolerance of protecting its senior management despite poor crisis management.

The Costa Cruise case was a mixture of unfortunate luck, poor onboard management and leadership in the grounding of the *Costa Concordia*, and lack of engine and power backup systems in the case of the *Costa Allegra*. The Costa Cruise Lines' mismanagement of crisis communications exacerbated a bad situation. When and if a CEO lacks the skills to communicate empathy and confidence, then the wisest action of management is to delegate this role to an executive who can be a credible face of the company.

The core risk management lesson from this chapter is: never take customers and stakeholders for granted. In tourism, customers have many choices of airlines, other modes of transport accommodation, destinations, attractions and restaurants. When customers receive value for money and excellent service, they let others know. When they are dissatisfied, they may not tell the service provider but they may tell their friends and relatives, or in some cases, millions of their contacts via social media.

The ultimate risk management practice is to ensure that all elements of the service chain within tourism businesses and destinations are working smoothly, and are overwhelmingly approved by customers and supported by stakeholders. In tourism, businesses should never let their guard down if they want to avoid an 'own goal' crisis.

Discussion Questions and Issues

1. You are the risk and crisis manager of a major tour operator; a group travelling on one of your European tour has been involved in a coach crash. One tour member was killed and ten injured. Discuss how you and your company will manage this unfortunate incident.

2. You are in charge of check-in for a major airline at a major airport. All passengers have turned up for the flight, meaning the flight is overbooked to the tune of four seats. How do you manage this situation in a manner that will retain the good name of the airline?
3. As the operations manager of a hotel, it has come to your attention that one of your staff members is insulting guests from China and this news has hit WeChat, the Chinese social media outlet to the tune of a million hits per day. Discuss how you are going to respond in a manner that will restore your hotel's reputation in China, and without sacking the staff member, who has a disability and valuable skills.
4. You are the shift manager of a budget hotel in London. A group of Muslim people has come for breakfast and the only menu option decided by the chef is bacon and eggs. He must be the only chef in the UK who doesn't realise that Muslims are forbidden to eat pork. Manage the situation.

REFERENCES

Air Travel Organiser's Licence (ATOL) (2020) www.caa.co.uk/atol-protection/. Accessed 3 December 2020.

BBC (2012) *Costa Allegra*. https://bbc.com/news/world-asia-17183134. Accessed 7 December 2020.

BBC (2015) *Costa Concordia*. https://bbc.com/news/world-europe-16563562. Accessed 7 December 2020.

BBC (2016) *Captain Francesco Schettino Costa Concordia*. https://bbc.com/news/world-europe-36421474. Accessed 7 December 2020.

BBC (2019a) *Thomas Cook: The Much Loved Brand with Humble Roots*. www.bbc.com/news/business-49789073. Accessed 27 November 2020.

BBC (2019b) *Thomas Cook Collapses As Last-ditch Rescue Talks Fail*, 23 September 2019. www.bbc.com/news/business-49791249. Accessed 7 December 2020.

Beirman, D. (2020) *The Implications of the 2019 Thomas Cook Collapse on Consumer Protection from Travel Business Insolvency in Australia and New Zealand*. Peer reviewed Working Paper, CAUTHE Conference February 2020: Auckland, New Zealand.

Benoit, W.L. (2018) 'Crisis and image repair at United Airlines. Fly the unfriendly skies', *Journal of International Crisis and Risk Communications Research*, 1(1): 11–26.

Canny, I. (2013) 'The role of food quality service and physical intertemporal consumer care environment on customer satisfaction and future behavioural intentions in casual dining restaurant', *Microeconomics: Intertemporal Consumer Choice and Savings E Journal*.

Carroll, D. (2009) 'United Breaks Guitars'. https://youtube.com/watch?v=5YGc4zOqozo. Accessed 3 December 2020.

Cordato, A. (2017) 'Dreamworld tragedy highlights legal duty of care for theme park guests', *Australian Leisure Managers.* https://ausleisure.com.au/news/dreamworld-tragedy-highlights-legal-duty-of-care-for-theme-park-guests/.

European Package Tour Directive (2020) https://ec.europa.eu/info/law/law-topic/consumers/travel-and-timeshare-law/package-travel-directive_en.

Hennig, S., Wasserburger, W., Wasserburger, M. and Satler, T. (2015) *How to Improve Accessibility in Natural Areas*. CORP Proceedings.

Laurence, R.D., Hong, S.J. and Cherrier, J. (2003) *Passenger Based Predictive Models of Airline No-show Rates*. SIGKDD, 24–27 August 2003, Washington, DC, USA.

Levin, B. (2017) 'United CEO offers worst possible response to airline's PR disaster', *Vanity Fair,* 11 April 2017. https://vanityfair.com/news/2017/04/united-airlines-oscar-munoz.

Liu-Lastres, B. and Johnson, A. (2019) 'Managing the reputation of cruise lines in times of crisis: A review of current practice', in Walters, G. and Mair, J. (eds), *Reputation and Image Recovery for the Tourism Industry*. pp. 179–94.

Merlin Entertainments (2020) www.merlinentertainments.biz. Accessed 4 December 2020.

Page, S. (2015) *Tourism Management*. Routledge: Oxon, Abingdon, UK. pp 296–317.

Shen, L. (2017) 'United Airlines stock drops $1.4 billion after passenger removal controversy', *Fortune,* 12 April 2017. https://fortune.com/2017/04/11/united-airlines-stock-drop/.

Smith, P. (2018) Personal communication. I have had many informal discussions with Peter Smith about the Swiss Canyoning disaster. He was the CEO of Contiki Australia in 1999 when the disaster occurred.

The Travel Corporation (2020) https://ttc.com.

United Airlines (2019) Annual Report. https://www.annualreports.com/HostedData/AnnualReportArchive/u/NYSE_UAL_2019.pdf. Accessed June 24, 2020.

Weaver, D. and Lawton, L. (2014) *Tourism Management* (fifth edition). John Wiley and Sons: Milton, Queensland, Australia. pp 58–9.

12

THEME 9: ENVIRONMENTAL CRISES AND RECOVERY

INTRODUCTION

Protection and preservation of the natural environment is one of the three pillars of sustainable tourism. Of those three pillars (social, economic and environmental), the impact of tourism on the environment has the longest-term consequences for the planet. Many tourism scholars have devoted their careers to researching the broad range of issues surrounding environmental sustainability (Harris et al., 2002; Holden, 2011; Wood, 2017). Scholars who specialise in environmental tourism hold a broad range of attitudes towards the relationship between environmental protection and tourism. A small number of tourism scholars (Gossling et al., 2010) advocate the reduction, or in extremis, the cessation of long-haul tourism to protect the global environment. There is a broad consensus among sustainable tourism scholars, policy makers and practitioners that the future of the planet and tourism depends on sustaining the Earth's environment. Stephen Wearing and Stephen Schweinsberg (2019) have provided excellent coverage of the ongoing debates among scholars who specialise in the environmental dimension of tourism. Ironically, during 2020, the Covid-19 pandemic led to a massive reduction in global tourism activity. Many destination regions concerned about over-tourism up until the end of 2019 became equally concerned about the absence of visitors during 2020. This highlighted the importance of striking a balance between environmental, social and economic sustainability. In 2020, many environmentally sensitive areas experienced an environmental resurgence and improvements to air and water quality due to the lack of human visitation during the pandemic.

At conferences and seminars where tourism scholars gather to share their research, environmental issues tend to dominate the agenda. Many tourism academics believe that they set the sustainability agenda in global tourism. However, all sectors of the travel industry have paid increasing attention to sustainability in general and environmental sustainability in particular. Certainly, from the travel industry's perspective, sustainability and eco-tourism sell. Sustainable tourism and eco-tourism are growing markets, which have attracted an increasing number of young and environmentally conscious travellers. Tourism businesses across all sectors of the tourism industry actively promote their 'green credentials', as do many destinations. The term 'green washing' applies to those marketing campaigns in which the companies 'stretch the truth' about their environmental credentials and in doing so risk exposure as frauds and the reputational damage that accompanies false marketing.

Environmental degradation and pollution, much of it caused by expanding human settlement, agricultural activity, industrial activity, mining, poorly planned tourism infrastructure and the harmful impacts of these activities on the environment, represent a very real risk for the environmental balance of the Earth and consequently for tourism.

There is a close link between the fate of tourism and the environment. Many of the world's most compelling tourism destinations and attractions are wonders of nature, due either to scenic grandeur or unique natural characteristics, including flora and fauna. Travellers are drawn to visit mountains, glaciers, polar ice, canyons, lakes, oceans, islands, beaches, coastal cliffs, bays, harbours, waterfalls, rivers, coral reefs, rainforests, deserts and desert scenery, wild animals, sea life in their natural habitat, volcanoes, craters, caves, taiga, autumn trees and sweeping grassland plains. These represent some of the many environmental treasures, which attract tourists to visit and experience the vast variety of unique natural attractions of planet Earth. Protection of the environment preserves ecosystems, vital to life on Earth. As human populations grow and human settlement occupies increasing areas of land, the habitats of thousands of endangered animal and plant species are at severe risk.

The relationship between tourism and the environment has varied enormously. At its best, tourism has been a force for protection of the environment. The establishment of national parks, from 1872, was intended to protect areas of natural beauty and biodiversity, and the flora and fauna extant within those areas. The establishment of protected natural areas was designed to benefit both the natural environment and ensure tourists could enjoy them for many generations. The world's first national park was Yellowstone National Park in the USA, established in 1872. The second was the Royal National Park, just south of Sydney, Australia in 1874. Since then, the establishment of thousands of national parks globally encompassed every form of terrain, habitat and climate on land and sea.

The negative dimension of tourism's interaction with the environment results from the fact that some environmental attractions are so popular that they attract large scale visitation, which frequently exceeds optimal carrying capacity in fragile environments. Carrying capacity is a core theme within environmental sustainability and some fragile environments can only bear a low level of visitation. Exceeding an optimal carrying capacity can impose negative impacts on the flora, fauna and water supplies of that area.

Massive and sometimes ugly tourism infrastructure developed close to many natural attractions catered to the perceived needs of visitors. There are countless examples to illustrate this tendency. Niagara Falls on the border of the US state of New York and the Canadian province of Ontario is a compelling example. The Niagara Falls are a spectacular natural site, but a short walk away from the falls is a Canadian tourism city with high rise hotels, cheap and tacky eateries, amusement centres, car and coach parking areas and a conurbation of urban ugliness, which surrounds this wonder of nature. The great challenge of sustainable environmental tourism is achieving a balance between the maintainance, protection and conservation of the environment, and ensuring that visitors can enjoy and experience environmental treasures while leaving the lightest possible footprint. Increasingly, the planning of environmentally sustainable tourism infrastructure involves a process which involves all three tiers of government (national, state/provincial and local), the local community and developers (Dredge and Jenkins, 2007).

Some famous beaches were originally pristine crescents of oceanic beach sand, prior to their discovery by tourists and tourism developers. Many of these beaches and their settings have been surrounded by high-rise hotels, apartments, shopping malls, entertainment precincts, freeways and airports. Waikiki (Hawaii), Copacabana (Brazil), Acapulco and Cancun (Mexico), Miami (USA), Surfers Paradise (Australia), the Costa del Sol and Majorca (Spain) are just a few examples where tourism development catering to the perceived needs and desires of tourists have despoiled places of natural beauty and created beachside tourism precincts, which are vulnerable to sea surges, hurricanes, typhoons, cyclones and tsunamis.

There is no shortage of case studies to illustrate environmental risks and crises in tourism. In this chapter, three broad case studies are discussed. One focuses on the environmental challenges facing national parks. A second focuses on the rising sea levels and the significant threat this poses to tourism and the very survival of low-lying island states such as the Maldives. A third case study focuses on the threat of rising water temperature and coral bleaching, which threatens the survival of Australia's Barrier Reef, the world's largest coral reef and the world's largest living organism. Threats to the Barrier Reef endanger a vast marine ecosystem dependent on the health of the coral reef.

In all the cases human activity, especially tourism, has been a contributing factor to the problem but effective risk and crisis management can be a contributor to addressing the problems. This chapter also addresses humanity's role in climate change and the impact of climate change on tourism.

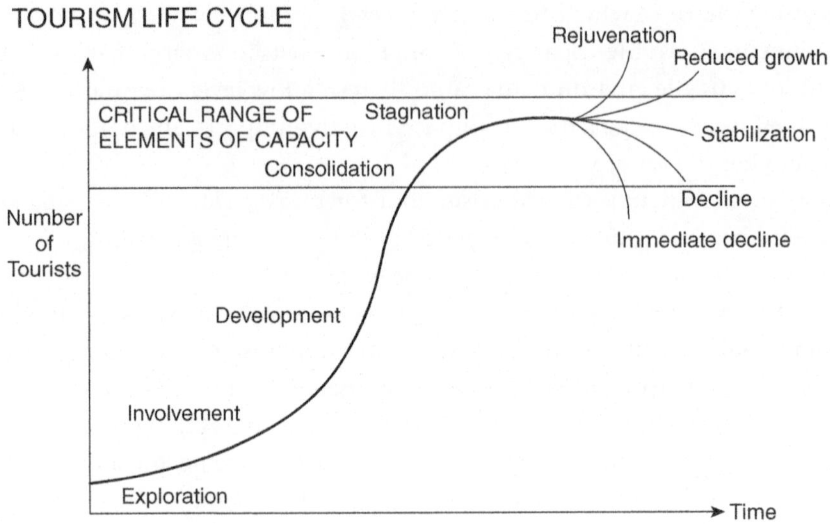

Figure 12.1 Butler's Tourism Area Life Cycle

Source: Richard Butler, 1980. Used with kind permission of Professor Richard Butler, Strathclyde University.

Of the many theoretical models which define the challenges for all tourism planners in balancing tourism needs and environmental protection, one especially relevant to this chapter is Richard Butler's Tourism Area Life Cycle model (Butler, 1980), also referred to as the Butler sequence (Weaver and Lawton, 2014). It makes the important point that most tourism destination attractions slowly move from having been 'discovered' to becoming integrated tourism attractions.

From an environmental risk management perspective, the process from exploration to consolidation needs to factor in a balance between access, number of tourists permitted at one time (carrying capacity) and the sustainable maintenance of the site, its ecosystem and environmental integrity. Success in achieving this balance will maintain the quality of the site and failure to do so will lead to a decline.

Environmental protection makes up a significant element in the United Nations Sustainable Development Goals (UNSDGs), as agreed by the UN General Assembly in 2015. Seven of the 17 Goals (see Figure 12.2) are relevant to environmental protection: 6, 7, 11, 12, 13, 14 and 15.

The UNSDGs have been widely embraced by the global tourism industry (in theory) via the leadership of the UN World Tourism Organization and it has been incorporated as policy by all key transnational, national and sectoral tourism industry bodies. Positive as this is, there are many tourism businesses and a few destinations that ignore these goals when it suits them and if the price is right. This is not a matter of being

Figure 12.2 United Nations Sustainable Development Goals 2015

Source: With kind permission United Nations. https://www.un.org/sutainabledevelopment/. The content of this publication has not been approved by the United Nations and does not reflect the views of its officials or Member States.

cynical, but regrettably, there are those elements within the tourism industry in which ethical conduct and concerns about environmental sustainability come a distant second to making millions of quick dollars.

Case Study 12.1

National Parks and World Heritage Sites

The core concept of national parks and protected areas involves controlled visitor access, the primacy of protecting flora, fauna and the physical environment within the park and minimising the human footprint on the park. National parks and World Heritage Sites represent markers for tourists that these areas are recognised as desirable and significant environmental and heritage attractions. In most cases, national parks are managed under the jurisdiction of national or provincial governments. As discussed earlier in this chapter, national parks have been in operation since 1872 but now exist in most countries in the world. In addition to the protection of sensitive natural environments, many national parks manage heritage sites, historically significant sites and unique cultural icons.

(Continued)

Since 1972 (100 years after the first national park opened), the UN, through the United Nations Educational, Scientific and Cultural Organization (UNESCO), sought to exert pressure on national governments to protect sites of unique and globally significant environmental and historical/heritage value by declaring them as UN World Heritage Sites (Batisse and Bollar, 2005). Many countries have supported the concept of world heritage listings as these sites have become marketable tourism drawcards. Unfortunately, there is no guarantee that UN World Heritage listing will result in governments implementing sustainable and environmentally sensitive management of these areas, although most do. One of the greatest environmental risks is that the infinite capacity for greed on the part of some individuals, corporations and governments can and often does result in unsustainable, commercial tourism development in and around a World Heritage Site, which despoils the site. In practice, World Heritage Sites only become properly governed as protected areas when a national or a regional government chooses to administer them.

Globally, the planning and operation of many (not all) national parks have been managed to prioritise the convenience of human visitors above protection of the parks' natural environment, flora and fauna. In order to enhance visitor access, many national parks are criss-crossed by heavily trafficked roads. These endanger animals traversing the park in search of food or seasonal migratory routes and tragically many of the parks' 'protected species' end their lives as roadkill. Most national parks are managed and funded by governments, although many national parks do charge an entry fee for day visitors.

In some safari national parks in Africa, the managers of national parks do the best they can to prevent the killing and abduction of animals in the park by professional poachers. However, many safari operators disrupt the hunting and natural activities of animals in order to show them off to tourists in jeeps who wish to photograph the 'big five'. The author was involved in a tourism recovery project in Kenya in 2003. After completion of the project, the team visited the Masai Mara National park, which adjoins the Serengeti National Park in Tanzania. In combination, this is one of the largest wildlife reserves in Africa – equivalent in area to twice the size of Wales. A common problem with wildlife reserves in Africa (and in other parts of the world) is that when one jeepload of tourists comes across a sleeping lion, word gets out and within ten minutes, ten jeeploads of camera-snapping tourists surround the lion, thus giving away the lion's location to all potential prey. Unsurprisingly, the lion wakes up in a foul mood, snarls at the cameras and misses a meal. This is one of countless examples of safari tour operators, with the best of intentions for their customers, disrupting the life cycles of animals and the parks that are intended to protect.

In most countries, national parks fall under multiple categories, three of which are outlined below (related to the ease or difficulty of access and recommended carrying capacity):

1. Recreation reserves are essentially green areas in or near urban areas, which are easily accessible to locals and visitors with minimal concern for the protection of fauna.

2. State or national parks have well-defined road or rail access to the park headquarters, which offers visitors a full range of facilities, including a restaurant or cafe and detailed information about the park. Vehicular traffic within the park is restricted to a small number of well-defined routes. These parks will also include a number of walking and bicycle trails. Many parks in alpine regions have overnight accommodation, much of it high standard, commercial accommodation operated in accordance to national park regulations for a controlled number of guests. Camping and built accommodation options in national parks are not confined to alpine ski resorts; they are quite common in national parks located far from major cities. Many national parks offer guided tours at a price.
3. Wilderness national parks are generally in remote regions and difficult to access, and are characterised by a very low carrying capacity. Visitors are required to be self-sufficient, remove all waste and bring their own food, water and supplies. Some may have transport access to the edge of the park but are required to traverse the park itself on foot. National park authorities have information in map, brochure or online form about permissible trails and highlights within the park.

National parks have unlimited potential to enhance the sustainability of the environment and ecosystem in natural tourist attractions but there are three caveats.

1. The quality and consistency of management and governance of national parks are critical. In practice, this depends on the relevant government's commitment to setting rules of engagement and enforcing them. In all likelihood, this will not be practised to a common standard in multiple jurisdictions in multiple countries. Consequently, some national parks in some parts of the world will be very successful in their mission to protect and preserve the natural environment and some will not. One of the biggest threats to national parks is the threat of natural disasters and forest/bushfires. Many national parks make inadequate preparation for such significant risks.
2. There are always questions about whether users/visitors to national parks respect the environment and the ecosystem within the park. Many do, but unfortunately a minority of tourists use national parks as dumping grounds, fail to extinguish barbeque fires, are careless with cigarette butts and interfere with both flora and fauna. Some also create danger for themselves and park staff when they explore beyond the designated walking routes, activating an expensive and time-consuming rescue. National park managers cannot eliminate the stupidity of some people but it's important that park managers ensure that visitors are made fully aware of their rights and responsibilities while in the park. The ultimate consequence of poor governance and uncontrolled behaviour of visitors is that the park, intended to protect an environment, can potentially destroy that environment.

(Continued)

3. Tourism companies operating within national parks are expected to respect and observe the rules of the park. While the needs of providing a memorable experience for tour groups is very important, it should not damage or disrupt the life of flora and fauna and the natural environment of the park. The top priority of national parks is the protection and preservation of a unique environment. Tour operators and guides can greatly enhance that visitor experience but not at the expense of nature.

ENVIRONMENTAL RISKS ARISING FROM TOURISM

Until the end of 2019, when tourism was at the height of its growth, many eco-destinations experienced over-tourism. This had negative social and economic consequences but also serious environmental consequences in many destinations. The sensitive environments of some destinations were literally being 'loved to death'. Some tourism development left a number of coastal and mountainous areas vulnerable to natural disasters, including sea surges, tsunamis and tropical storms. This was due to the removal of natural protections such as mangroves in seaside locations or trees in mountain locations. During the Indian Ocean tsunami of December 2004, the destruction of mangroves in Phuket enabled the tsunami wave to rush onto land unimpeded by natural defences. The removal of vegetation cover has facilitated mudslides in some mountain locations. Since the 2004 Indian Ocean tsunami, many countries most severely affected, including Thailand, Sri Lanka, India and Indonesia, have replanted the mangrove swamps in coastal areas affected by inundation.

In 2019, global attention was drawn by extensive media coverage to lengthy queues of mountaineers attempting to scale the world's highest peak, Mount Everest, bordering Nepal and Tibet. Ascending the peak has always been exceedingly dangerous. The lengthy exposure of climbers to the severe and changeable conditions on the slopes of Mount Everest led to the death of 11 climbers and support staff (out of 1,100) during the April–May 2019 climbing season (*The Guardian*). In 2019, queues of climbers were exposed to the changeable weather for many hours while attempting to scale the summit. The steep slopes of the mountain are littered with frozen bodies and detritus, left by hundreds of climbers who failed or succeeded in the attempt. Climbers and their sponsors are prepared to pay support teams large sums of money (on average US$11,000) to assist them in summiting Everest.

While tourism can be a positive force for protecting parts of the global environment, if tourism is not managed sustainably, it can also be a force for environmental degradation and pollution. Transport on land, sea and air contributes over ten per cent of global carbon emissions. Mass tourism in many popular destinations consumes vast amounts of water (far more than local residents), which threatens the water table in many destinations. Bali is one of many examples, despite a high level of rainfall. Mass tourism adds to the sewage waste and garbage generated by high levels of over-packaged, fast-food consumption and litter. In many popular tourism destinations, tourists create more pollution than locals.

CLIMATE CHANGE AND ITS IMPACT ON THE ENVIRONMENT AND TOURISM

Many of the discussions about climate change in the political, media and sometimes academic discourses are simplistic to the point of being an insult to intelligence. In some academic and political circles, anyone labelled as a climate change sceptic is regarded a deranged lunatic. Many observers of climate change orthodoxy are on the bandwagon of political correctness. Sadly, the orthodoxy tends to take the limited view that climate change is overwhelmingly the fault of Western society.

Climate change on Earth has been an unending process since the planet coalesced out of inter-stellar matter 4.5 billion years ago. Earth's atmosphere has changed massively over a eons, and oceans have formed. The eruption of massive volcanoes led to extended periods of extreme cold. The most recent such event followed the eruption of the Tambora volcano, Indonesia in 1815. The planet has experienced heat waves and ice ages. Sea levels have risen and fallen. Collisions with comets and asteroids have wiped out millions of species (extinction events), which is why one can only see dinosaurs as fossils or skeletons in museums. All this occurred before Homo Erectus (humanity's ancestor) appeared on the Earth, at most a mere two million years ago. The activity of the sun, changes in our orbit around it and global climatic patters occur independently of any human influence.

The world has been industrialised for just over 250 years. The above points are for the sake of context. Industrialised humanity and burning of fossil fuel for the benefit of today's eight billion human beings have indeed contributed to the current phenomenon of climate change. That is not in dispute. However, the arguments advanced by some that climate change is a purely human-based phenomenon lack any credibility. Scientists can postulate and argue about the extent of human influence on climate change. However, in the context of a small planet in a small solar system on the edge of a galaxy with 100 billion other stars, which is one of millions of other galaxies, we are a lot less influential in altering the climate of our planet, let alone the cosmos, than we like to imagine.

Arguably, one of the greatest environmental threats to tourism attractions and destination environments is related to climate change. There is a legitimate debate on the extent to which climate change and its consequences arise from natural forces or human activities. Nevertheless, the overwhelming scientific evidence reveals that climate change has occurred at an accelerated rate in recent years. Ice caps in both the northern polar regions and the southern polar regions are melting at an alarming rate and sea levels are rising. In some regions of the world, water and air temperatures have risen considerably in recent years. The consequences have included:

- Extensive coral bleaching in the Great Barrier Reef and other coral reefs.
- Shortened ski seasons and lack of snow in marginal ski resort areas.
- Changed courses of hurricanes, cyclones, typhoons and tornadoes threatening localities not previously hit by such weather events.
- Raised temperatures that have increased the likelihood, severity and extent of bush and forest fires on both the natural environment and settled communities.
- Air and water pollution arising from industrial and vehicle emissions in the cities of Europe, North America, the Indian subcontinent, East and South-East Asia.
- Inundation of some coastal areas as a result of rising sea levels and higher sea surges.
- Changes in weather patterns resulting in extreme episodes of droughts and floods.
- Deforestation in the Amazon and South-East Asia, which alters climate in affected regions.
- The likely inundation of low-lying islands states, including the Maldives in the Indian Ocean and the South Pacific island states of Kiribati, Marshall Islands and Tuvalu.
- Melting of glaciers and the associated threat of mudslides in mountainous regions.

Tourism has been one of many contributing factors to environmental degradation. This has resulted in the increased vulnerability of many popular tourism destinations to natural disasters. Climate changes have altered the paths of many hurricanes. This has meant that countries formerly not on the traditional hurricane path are now vulnerable. Jamaican meteorologist Franklin McDonald (2005) made a presentation to the June 2005 Best Education Network in Kingston, Jamaica. He predicted that based on 20 years of hurricane movements, hurricane paths due to rising sea temperatures in the Caribbean Sea were moving west. During that presentation, he predicted that both Jamaica and New Orleans were now firmly on the hurricane path. Just over a month after his presentation, Hurricane Katrina devastated New Orleans in August 2005.

Human encroachment into the natural environment has heightened the risk of exposure to bush and forest fires. This was at its most evident in South-East Australia in late 2019 and early 2020, and in California between August and September 2020. Many of the places affected by the fires were popular tourism destinations and in both cases many visitors were trapped in communities and beaches ringed by fire. Annual crop burnings in Indonesia, Myanmar, Malaysia and northern Thailand have resulted in poor air quality over much of South-East Asia for the period May–September for most of the past 20 years. Humanity can control its contribution to climate change, but it cannot stop climate change arising from natural sources.

During 2018–2019, Swedish teenager Greta Thunberg attracted extensive coverage in the global media and a following among young people for her angry and impassioned denunciation of world leaders for their alleged lack of recognition and concern for the impacts of climate change. Although her criticisms were not original, coming as they did from an eloquent teenager who focused on the harm climate change would impose on future generations, she gained considerable global attention (BBC, 2020).

One of the few positive outcomes of the Covid-19 pandemic (discussed in Chapter 3) has been the temporary relief of environmental degradation due to tourism. The curtailment of international tourism movements has contributed to a reduction of the human footprint on many natural environments and a significant reduction in transport-based pollution.

Case Study 12.2

The Maldives and Rising Sea Levels

There is a standing joke in the Maldives that if you are taking a walk down the main street of the capital city, Male, it's a good idea to take flippers, a mask and a snorkel with you just in case the big wave hits.

The Maldives is a cluster of low-lying coral atolls in the Indian Ocean just north of the Equator and a few hundred kilometres south-west of southern India. At its highest point, the Maldives is just under three metres (ten feet) above sea. The Maldives is a country with a high level of economic dependence on tourism, which in 2019 comprised 30 per cent of its GDP and 40 per cent of its tax revenue.

Professor Jack Carlsen (2009) has been among the most active researchers on crises affecting the Maldives. During the Indian Ocean tsunami on 26 December 2004, there were 82 lives lost, 26 missing and over 100,000 rendered homeless in the Maldives. The airport was under water for several hours, which hampered evacuation and rescue and relief. As was the case in Thailand, thousands of tourists, most of them from Europe, were caught up during the tsunami. Of the country's 87 resorts, 19 were forced to close after having sustained severe damage. The key source markets from the Maldives at that time were UK, Italy, Germany, Japan and France. The tsunami was a major crisis for the Maldives as a nation and as a tourism destination. International tourism arrivals in 2005 were 40 per cent down on the levels of 2004 (Carlsen, 2009).

The government of the Maldives is painfully aware of its vulnerability to sea level rises and is constantly working to promote tourism as its only viable source of income.

(Continued)

The country lacks the land to grow food and manufacture goods. Its only natural resource is fish, but mass fishing would detract from the main reason tourists visit the Maldives, the experience of a coral wonderland full of tropical marine life.

After the tsunami, the Maldives government developed a tourism master plan between 2007 and 2011 to expedite tourist evacuation, and the establishment of a Tourism Emergency Operation Centre in the event of a future inundation event, which could coordinate rescue, medical assistance and food aid.

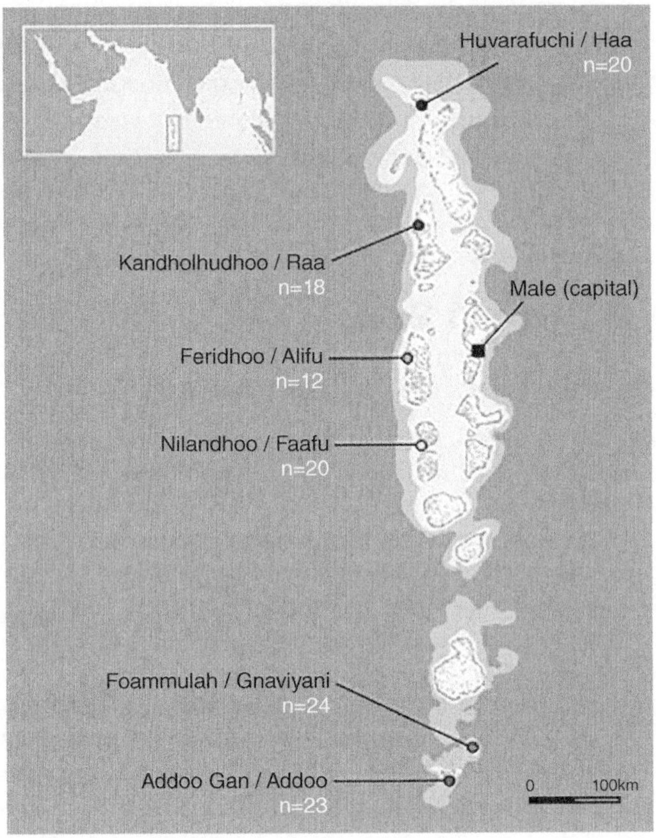

Figure 12.3 Map of the Maldives

Source: Reprinted with permission from Encyclopadia Britannica, © 2002 by Encyclopadia Britannica, Inc.

During 2020, the Maldives, in common with all tourism destinations, was severely affected by the Covid-19 Pandemic. The country is so desperate to maintain tourism income that since July, it has declared the Maldives open for tourism from all countries free of quarantine restrictions, provided visitors have evidence of a negative test result prior to arrival. However, the hope for tourism influx has failed to materialise due to restrictions and lockdowns in key source markets, a severe lack of flights and the requirements of most source market countries that travellers have to quarantine on return.

Pandemics may come and go, but for the Maldives the potential crisis that remains top of mind is the sea level and its impact of a sea level rise on this precariously located nation.

RECOVERY FROM ENVIRONMENTAL CRISES

The recovery of tourism businesses and destinations from environmental crises should be part of a collaborative approach in which tourism businesses and destinations work with government, emergency management agencies and other businesses. Tourism sectors have paid close attention to adapting to technologies which reduce environmental impacts. Airlines have adopted alternative fuels to aviation fuel and flight routes that reduce fuel consumption. The Boeing 787 Dreamliner, in much of its fuselage and wing construction, is constructed of carbon fibre, which reduces the weight of the aircraft which in turn reduces fuel consumption by 20 per cent compared to previous aircraft. There is a strong element of self-interest in the cost savings for airlines from reduced fuel consumption although they benefit the environment. Cruise ships have paid a great deal of attention to reducing their environmental footprint via a number of measures ranging from more fuel-efficient engines, use of wind power on a number of cruise ships, and strict protocols enforced by the International Maritime Organisation (IMO) for waste and sewage disposal and water recycling.

The accommodation sector actively promotes a range of green initiatives, which include dual flush toilets, LED lighting, water savings through water-efficient shower heads and timers on showers and electrical appliances. Many hotels in central business districts (CBDs) have introduced herb and salad gardens on the roofs of high-rise hotels and vertical gardens. In environmentally sensitive areas, glamping and eco-lodges have been designed to minimise the environmental footprint of hotel accommodation without compromising on comfort. Hotels have played a significant role in accommodating emergency management personnel and facilitating communication for fire-fighting units and rescue personnel during cyclones and bushfires in Australia and hurricanes in the South-West Pacific.

Tourism businesses are increasingly linked into emergency management communication, and their equipment and infrastructure can be of great assistance during crisis events that are related to environmental issues.

Case Study 12.3

The Great Barrier Reef, Australia – A Sensitive Environment Under Threat

The Great Barrier Reef is one of the natural wonders of the world. Naturalist and renowned wildlife documentary maker Sir David Attenborough has done more than any person on Earth to raise consciousness of the Earth's environmental challenges. He declared that while the Great Barrier Reef is his favourite place in the world to visit, he is deeply concerned about its long-term survival prospects due to climate change, and to a lesser extent, pollution.

The Great Barrier Reef is the world's largest living organism and the only living thing on Earth visible from space. The reef is made up of billions of coral polyps. Over millions of years, they have built a reef of coral cays, atolls and islands, which stretch off the Queensland coast in Australia from the industrial city of Gladstone just south of the Tropic of Capricorn to the southern coast of Papua New Guinea. The reef is 2,100 kilometres (or 1,250 miles) in length.

Coral reefs are a fragile environment. Much of the Great Barrier Reef is between 40 and 60 kms offshore, although the reef comes close to the coast at the popular tourism centres of Cairns and Port Douglas in far North Queensland. A number of islands off the Queensland coast accessible from Rockhampton, Airlie Beach, Townsville, Mission Beach and Cairns have tourism resort developments. The Great Barrier Reef has contended with industrial and mining pollution in its southern extremity near Gladstone. From the ports of Gladstone, Townsville and Mackay, Australian shipments of alumina, aluminium, coal, gas, minerals, live sheep and cattle, and agricultural produce navigate the reef to export markets in Asia and the rest of the world, occasionally damaging parts of the reef when they deviate from the correct course. Tourism activity also makes a small addition to pollution in the reef.

Although mining and industry are required by law to avoid polluting the waters off Queensland's coast, breaches have occurred. The reef faces three key environmental challenges:

1. Pollution and man-made damage to coral.
2. The crown-of-thorns starfish, which feeds on coral and has destroyed vast areas of the reef.
3. Warming of the sea above tolerable temperatures, which creates coral bleaching.

The third of these presents the most serious threat to the survival of the reef. Coral has a minimal tolerance for variations in water temperature. An increase of one to two degrees Celsius can trigger coral bleaching and a cessation of the breeding cycle of coral polyps. This in turn drives away the complex ecosystem, which operates on and around a healthy and living coral reef.

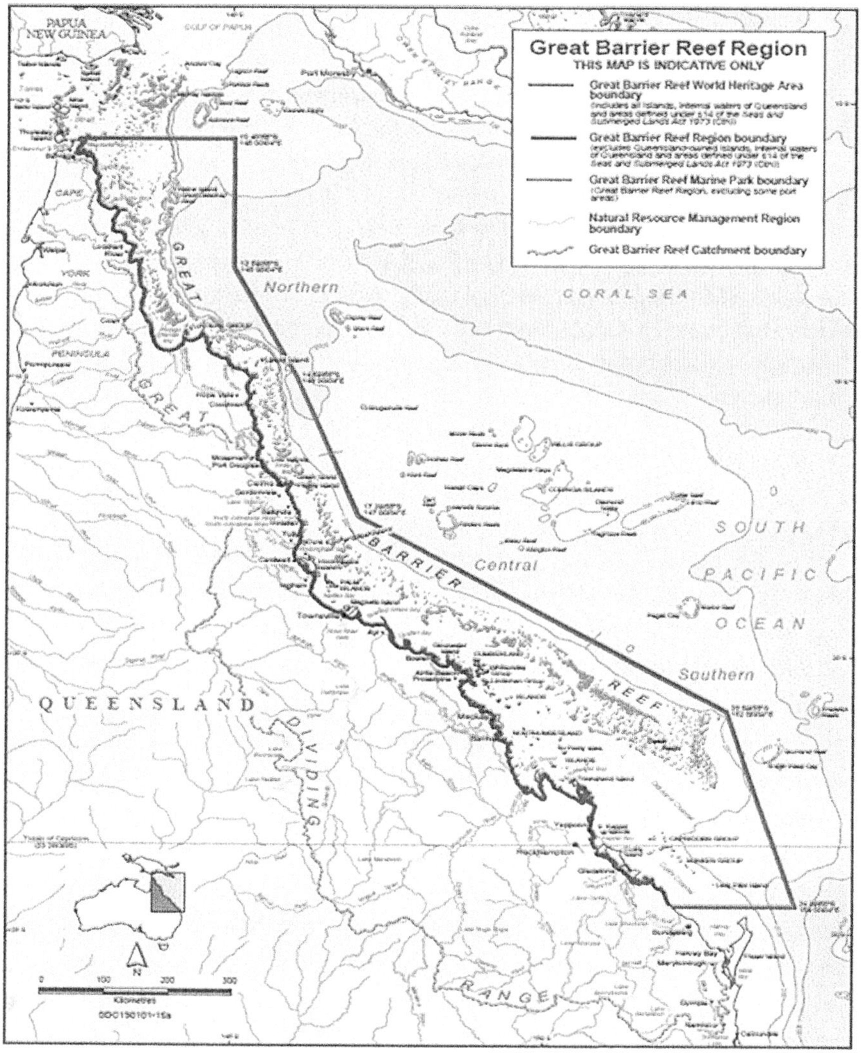

Figure 12.4 The Great Barrier Reef, Queensland, Australia

Source: Great Barrier Marine Park Authority 2021: https://www.gbrmpa.gov.au. NB The administration of the Great Barrier Reef Marine Park Authority has been under control of the Australian Government since 1975. It is Australia's largest world heritage listed sites.

By 2020, scientists believed that close to 90 per cent of the Great Barrier Reef was affected by coral bleaching. From a tourism perspective, the Great Barrier Reef and the

(Continued)

extensive ecosystem of marine life that it supports are the core attraction to hundreds of thousands of domestic and international tourists who visit North Queensland. In 2020, interstate domestic tourism and international tourism to the Barrier Reef was disrupted by Covid-19 and the Queensland government's resistance to non-Queenslanders visiting the state, but even before the pandemic there was considerable concern that the state of the reef could reduce its allure as a tourism destination.

In Queensland, Cairns-based James Cook University's Centre for Disaster Management and scholars at Central Queensland University, which has campuses in Cairns, Townsville and other major coastal cities in Queensland, are paying an active multidisciplinary role in assessing the extent of the threat to the reef and strategies to counter it. Queensland has long been a centre of tourism for Australian and international visitors and the Great Barrier Reef is the core attraction, especially for international visitors.

SUMMARY AND CONCLUSION

The global environment is tourism's greatest asset and a primary attraction for tourists. Travellers undertake travel for a variety of reasons, but the majority of leisure travellers aspire to experience at least some of the natural wonders of the world. Dreaming of majestic mountains, glaciers, pristine beaches, coral wonderlands, dramatic desert scenery, Africa's wildlife or tropical rainforest country in Brazil or Borneo is central to the tourism experience. Tourists want to experience these places in all their natural purity. Responsible eco tourists will visit, savour the grandeur, commit it to photos and videos, and leave it as found. At its core, sustainable tourism affirms that the descendants of today's tourists, several generations hence, can enjoy the same experience.

There is a growing awareness among tourism scholars and tourism practitioners that the demand for environmentally sustainable tourism and planning approaches for destinations is growing. Sustainability is the ultimate multidisciplinary area of tourism research, incorporating all the sciences, law and ethics, politics, geography and climatology. Unfortunately, selfish attitudes held by some travellers and tourism entrepreneurs remain. 'We pay our money and we want our money's worth' is common justification for fouling the environment. This self-centred, 'live for today' attitude has transformed formerly pristine beaches into strips of high-rise hotels, apartments and offices built on sand spits, which could potentially become a death trap during a sea surge in Miami, Cancun, Copacabana or Surfers Paradise.

National parks and World Heritage Sites contribute to the protection of the environment when managed in a manner that prioritises the environment and the ecosystem ahead of the visitor. That may sound unfair to the paying tourist, but the tourist is a visitor into nature's realm and is required to be a considerate guest.

Humanity's impact on planet Earth has been profound and often malign. While humans are far from the sole cause of climate change, there is little doubt that our profligate exploitation of the Earth's resources has contributed to a wide range of environmental problems and exacerbated the natural processes of climate changes. Tourism alone cannot solve all environmental risks and problems, but tourism can make a shift from being part of the problem to playing an integral role in working towards a sustainable solution.

Discussion Questions

1. You are planning an eco lodge in a destination of your choice. Discuss some of the measures you would incorporate to ensure that it was environmentally sustainable while maintaining a high level of comfort for guests.
2. You are in charge of a wildlife national park in Africa. Discuss some of the ground rules you would develop for safari tour operators who take tourists into your park.
3. If you were placed in charge of a national park in an alpine area, how would you plan the development of walking trails and ski runs?
4. If you were planning to develop a historic walking path in a country of your choice, how would you make it relevant to visitors from many countries?

REFERENCES

Batisse, M. and Bollar, G. (2005) *The Invention of World Heritage*. History Club Association of Former UNESCO Staff Members, UNESCO.

BBC (2020) *Greta Thunberg: Who is She and What Does she Want?* https://www.bbc.com/news/world-europe-49918719. Accessed 20 Dec 2020.

Butler, R. (1980) 'The concept of the tourism area life cycle', *Canadian Geographer*, 24(1): 5–12. There have been countless papers and books written by Richard Butler and others which discuss Butler's TALC.

Carlsen, J. (2009) 'Disaster response and tourism recovery strategies in the Maldives', in Pforr, C. and Hosie, P. (eds), *Crisis Management in the Tourism Industry. Beating the Odds*. Ashgate: Farnham, Surrey, UK. pp. 131–50.

Dredge, D. and Jenkins, J. (2007) *Tourism Planning and Policy*. John Wiley and Sons: Milton, Queensland, Australia. pp. 106–33.

Gossling, S., Hall, C.M., Peeters, P. and Scott, D. (2010) 'The future of tourism: can tourism growth and climate policy be reconciled? A climate change mitigation perspective', *Tourism Recreation Research*, 35(2): 119–30.

Harris, R., Griffin, T. and Williams, P. (2002) *Sustainable Tourism. A Global Perspective.* Elsevier: Oxford, UK.

Holden, A. (2011) *Environment and Tourism* (second edition). Routledge: Abingdon, Oxon, UK.

McDonald, F. (2005) *Presentation on the Meteorological Changes and Altered Hurricane Paths*, Best Education Network, Think Tank 3: Kingston, Jamaica.

The Guardian (2019) Congestion on Mount Everest leads to Backlog of Climbers in 'Death Zone'. https://www.theguardian.com/world/2019/may/23/congestion-mount-everest-backlog-climbers-death-zone. Accessed 18 May 2021.

Tourism Queensland (2020) http://www.queensland.com. Accessed 18 May 2020.

United Nations (UN) (2015) *UN Sustainability Goals*. https://sdgs.un.org/goals. Accessed 4 December 2020.

Wearing, S. and Schweinsberg, S. (2019) *Ecotourism: Transitioning to the 22nd Century.* Routledge: Abingdon, Oxford, UK.

Weaver, D. and Lawton, L. (2014) *Tourism Management* (fifth edition). John Wiley and Sons: Milton, Queensland, Australia. pp. 287–94.

Wood, M. (2017) *Sustainable Tourism on a Finite Planet: Environmental, Business and Policy Solutions*. Routledge: Abingdon, Oxon, UK.

INDEX